An Introduction to Agro Forestry

An Introduction to Agro Forestry

Amalendu Bhattacharya

RANDOM PUBLICATIONS

NEW DELHI - 110 002 (INDIA)

An Introduction to Agro Forestry

ISBN 978-93-51112-74-7

Published in 2014 in India by

RANDOM PUBLICATIONS

4376-A/4B, Gali Murari Lal, Ansari Road
New Delhi-110 002
Phone: +9111-43580356, 23289044
E-mail: randomexports@gmail.com; sales@randompublications.com;
info@randompublications.com

Reprinted 2026

Type Setting by: Friends Media, Delhi-110089

Preface

Agroforestry is an integrated approach of using the interactive benefits from combining trees and shrubs with crops and/or livestock. It combines agricultural and forestry technologies to create more diverse, productive, profitable, healthy, and sustainable land-use systems. A narrow definition of agroforestry is "trees on farms." The theoretical base for agroforestry comes from ecology, via agroecology. From this perspective, agroforestry is one of the three principal land-use sciences. The other two are agriculture andforestry. The efficiency of photosynthesis drops off with increasing light intensity, and the rate of photosynthesis hardly increases once the light intensity is over about one tenth that of direct overhead sun. This means that plants under trees can still grow well even though they get less light. By having more than one level of vegetation, it is possible to get more photosynthesis than with a single layer. Agroforestry has a lot in common with intercropping. Both have two or more plant species (such as nitrogen-fixing plants) in close interaction, both provide multiple outputs, as a consequence, higher overall yields and, because a single application or input is shared, costs are reduced. Beyond these, there are gains specific to agroforestry. Agroforestry is also defined as the deliberate land management unit whereby, woody perennials are deliberately grown on same piece of land along with agricultural crop and or livestock in some form of spatial arrangement or temporal sequence.

There is some evidence that, especially in recent years, poor smallholder farmers are turning to agroforestry as a mean to adapt to the impacts of climate change. A study from the CGIAR research program on Climate Change, Agriculture and Food Security (CCAFS) found from a survey of over 700 households in East Africa that at least 50% of those households had begun planting trees on their farms in a change from their practices 10 years ago. The trees ameliorate the effects of climate change by helping to stabilize erosion, improving water and soil quality and providing yields of fruit, tea, coffee, oil, fodder and medicinal products in addition to their usual harvest. Agroforestry was one of the most widely adopted adaptation strategies in the study, along with the use of improved crop varieties and intercropping. Groforestry represents a wide diversity in application and in practice. One listing includes over 50 distinct uses. The 50 or so applications can be roughly classified under a few broad headings. There are visual similarities between practices in different categories. This is expected as categorization is based around the problems addressed and the overall economic constraints and objectives.

This book describes the new imaging techniques being developed to monitor this subjects.

I thank all members of my team who have helped in the preparation of the book. My special thanks go to "Random Publications" who have published the book.

—Amalendu Bhattacharya

Contents

1

Sustaining the Multiple Functions of Agroforestry Biodiversity

Introduction

Throughout the world human communities have played a central role in shaping nature's diversity and its associated functions. Biological diversity has contributed in many ways to the development of human culture, and mankind has in turn influenced biological diversity. As an example, gourds (*Lagenaria siceraria* and *Luffa spp.*) show tremendous varietal diversity and have been selected for a multitude of use and functions, including containers, pipes, scrubbers, floats, musical instruments, penis sheaths, ornaments and food. Plants and animals, both wild and cultivated, have been combined in complex and diverse agroecosystems in terrestrial and aquatic environments. At the broader landscape level, recent scientific evidence suggests that virtually every part of the globe-from boreal forests to the humid tropics-has been inhabited, modified and managed for millennia. Over time, human influence has shaped the expression of agricultural biodiversity at the genetic, species, ecosystem and landscape levels.

The present paper is a contribution to the FAO/Netherlands Conference on the "Multifunctional Character of Agriculture and Land-MFCAL" that will take place in Maastricht, The Netherlands, in September 1999. This background paper focuses on the relationships between agricultural biodiversity and the functions of agro-ecosystems at different spatial and temporal scales. Selected examples are used to highlight the multiple functions of agricultural biodiversity and its links with rural livelihoods in a range of ecological and economic settings. The paper is divided into three parts. Part one draws from the MFCAL approach, and in particular focuses on the contribution made by agricultural biodiversity to 1) food and livelihood security 2) production and environmental sustainability and 3) rural development. Part two identifies the forces which impact on agricultural biodiversity. Part three concludes by outlining some of the policy and institutional reforms needed to sustain agricultural biodiversity and agro-ecosystem functions.

The MFCAL Approach

The MFCAL Approach provides the overall referential framework for this paper. The approach recognises that the first and foremost role of agriculture remains the production of food. It stresses however, that agriculture and related land use activity can also deliver a wide range of non-food goods and services, influence the natural resource system, shape social and cultural systems and can contribute significantly to economic growth. The approach also focuses on the trade-offs and synergies that can exist between these different functions.

The first dimension of the multifunctional character of agriculture and related land use concerns food production and the contribution that this makes to food security.

Food security has been defined by FAO as a situation in which all people at all times have physical and economic access to sufficient, safe and nutritious food to meet their dietary needs and food preferences for an active and healthy life. There are three dimensions implicit in this definition: availability, stability and access.

Adequate food availability means that, on average, sufficient food supplies should be available to meet consumption needs. Stability refers to minimising the probability that, in difficult years or seasons, food consumption might fall below consumption requirements. Access to food draws attention to the fact that, even with bountiful supplies, many people still go hungry because they are poor and unable to produce or purchase the food they need. In addition if food needs are met through exploiting non-renewable natural resources or degrading the environment there is no guarantee of food security in the longer-term.

Current concerns for food security stem from both the unacceptability of current levels of food insecure people (at least 800million people) and the recognition that agriculture will have to feed an increasing human population, forecast to reach 8 000 million by 2020, of whom 6 700 million will be in developing countries. In most developing countries, the majority of the poor live in rural areas and depend on agriculture for their livelihoods. Expanding food production to feed this increasing population, while alleviating poverty through gainful employment in agriculture, is a formidable challenge.

In addition to food production and the vital contribution that it makes to food security, the MFCAL Approach recognises three further broad functions 1) environmental 2) economic 3) social.

The Environmental Function. Agriculture and related land use can have beneficial or harmful effects on the environment. The MFCAL approach can help to identify opportunities to optimise the linkages between agriculture and the biological and physical properties of the natural environment. The environmental function of the MFCAL Approach is relevant to a number of critical global environmental problems including biodiversity, climate change, desertification, water quality and availability, and pollution.

The Economic Function. Agriculture remains a principal force in sustaining the operation and growth of the whole economy, even in highly industrialised countries. Valuation of the various economic functions requires assessment of short, medium and long-term benefits. Important determinants of the economic function include the complexity and maturity of market development and the level of institutional development.

The Social Function. Both the maintenance and continuing dynamism of rural communities are basic to sustaining agro-ecology and improving the quality of life (and indeed, to assuring the very survival) of rural residents, particularly of the young. On another level, the capitalisation of local knowledge and the forging of relationships between local and external sources of expertise, information and advice are fundamental to the future of existing rural communities. Social viability includes maintenance of cultural heritage: for we know that in many instances, societies still identify strongly with their historical origins in agrarian communities and rural lifestyles.

The Multiple Functions of Agricultural Biodiversity

Agro-ecosystem functions are partly determined by the social goals of farmers, pastoralists, forest dwellers, fisherfolk and gardeners,- men, women and children with their own definitions of well being and their different priorities, rights, capabilities and knowledge.

These social goals include economic, cultural and often aesthetic values as well as those of biological production. Depending on circumstances, preference may be given to short term maximisation of specialised productivity based on a single crop or to the diversity and persistence of production. These factors influence the way in which biodiversity is managed from the level of a discrete production unit (pond, field, swidden garden) right up to the larger landscape that is continuously transformed through the interplay between human agency and ecological processes (e.g. forests, pastoral landscapes, coastal zones and mangroves).

Both natural processes and human management have generated and sustained a vast array of genetic, species and ecological diversity. In turn, this agricultural biodiversity performs many different socioeconomic and environmental functions which are closely interrelated. Given the strong historical link between rural livelihoods and the components of biodiversity that are managed in different ways for different purposes, agro-ecosystems necessarily include by definition people and their institutions as well as the agricultural biodiversity that they co-create and use.

This inclusive view is implicit in the concepts and definitions jointly developed by the FAO and the Secretariat of the Convention on Biological Diversity. Considering agricultural biodiversity through such a holistic framework encourages the kind of methodological pluralism that is key to understanding the structure and functions of agricultural biodiversity in time and space.

Key Concepts and Definitions

Agricultural biodiversity or agrobiodiversity. Agricultural biodiversity refers to the variety and variability of animals, plants, and micro-organisms on earth that are important to food and agriculture which result from the interaction between the environment, genetic resources and the management systems and practices used by people. It takes into account not only genetic, species and agro-ecosystem diversity and the different ways land and water resources are used for production, but also cultural diversity, which influences human interactions at all levels. It has spatial, temporal and scale dimensions. It comprises the diversity of genetic resources (varieties, breeds, etc.) and species used directly or indirectly for food and agriculture (including, in the FAO definition, crops, livestock, forestry and fisheries) for the production of food, fodder, fibre, fuel and pharmaceuticals, the diversity of species that support production (soil biota, pollinators, predators, etc.) and those in the wider environment that support agro-ecosystems (agricultural, pastoral, forest and aquatic), as well as the diversity of the agro-ecosystems themselves.

Agricultural ecosystems or agro-ecosystems. Agro-ecosystems are those "ecosystems that are used for agriculture" in similar ways, with similar components, similar interactions and functions. Agro-ecosystems comprise polycultures, monocultures, and mixed systems, including crop-livestock systems (rice-fish), agroforestry, agro-silvo-pastoral systems, aquaculture as well as rangelands, pastures and fallow lands. Their interactions with human activities, including socioeconomic activity and sociocultural diversity, are determinant.

Agro-ecosystems may be identified at different levels or scales, for instance, a field/crop/herd/pond, a farming system, a land-use system or a watershed. These can be aggregated to form a hierarchy of agro-ecosystems.

Ecological processes can also be identified at different levels and scales. Valuable ecological processes that result from the interactions between species and between species and the environment include, inter alia, biochemical recycling, the maintenance of soil fertility and water quality and climate regulation (e.g. micro-climates caused by different types and density of vegetation).

Moreover, the interaction between the environment, genetic resources and management practices influence the evolutionary process which may involve, for instance, introgression from wild relatives, hybridisation between cultivars, mutations, and natural and human selections. These result in genetic material (landraces or animal breeds) that is well adapted to the local abiotic and biotic environmental variation.

Agricultural Biodiversity's Contributions to Food and Livelihood Security

Livelihood systems are diverse in rural areas and vary among different cultural groups and in different regions of the world. They commonly rely on a mix of wild foods, agricultural produce, remittances, trading and wage labour. Empirical evidence from many different locations suggests that rural households do engage in multiple activities and rely on diversified income portfolios. Contrary to received wisdom, the actual contribution of agriculture to livelihoods can be quite low in today's fast changing rural areas.

In sub Saharan Africa, for example, a range of 30-50 % reliance on non-farm income sources is common but it may reach 80-90% in Southern Africa (Ellis, 1999). Household decision making continually adjusts to the changing nature of the environment, local economies and governance. At higher levels, it is simply impossible to predict the relationships between agro-ecosystems and households, particularly in resource-poor areas where there is much biological and social diversity.

The tendency for rural households to engage in multiple occupations is often mentioned, but few attempts have been made to link this behaviour in a systematic way to agricultural biodiversity and its multiple functions. In reflection of sectoral interests and disciplinary specialisations, the conventional point of entry for scientific research, management and policy has been to focus on selected components of agricultural biodiversity (e.g. plant genetic resources). However, this approach often leads to a mismatch between standard development interventions and diverse local realities, needs and priorities. Reversing this approach requires putting people with their assets, activities, and complex livelihoods at the centre of analysis (Chambers, 1997). The functions of agricultural biodiversity thus need to be situated and mapped out within a total livelihood context.

Dynamic and complex livelihoods usually rely on plant and animal diversity, both wild and in different stages of domestication.

Agricultural Biodiversity Meeting Human Needs

- There is high intra-specific genetic variation in the date palm oasis agro-ecosystems of Algeria, Chad and Egypt (Barakat, 1995). The principal varieties differ from one oasis to another. In general, there are more than ten varieties of date palms in each oasis. In a well organised and maintained palm grove, the owner plants different varieties of dry and semi-dry dates that mature in different months to meet the demands of local consumption and the market. Moreover, each genetic variety confers its own unique stamp on i) the taste of the date fruit and the wine made from it ii) the texture of the edible palm centre iii) the properties of the wood from the palm trunk that is used in construction and tool making iv) the mechanical qualities of palm leaves used for ceilings and fences as well as those of the palm's fibrous leaf base used to make ropes and sacks and v) the nutritional values of date stones fed to camels.

- About 25% of the land in the small holding sector are under home gardens in Sri Lanka. The garden system is similar to the complex structure and multiple functions of the forest, though not identical to it. Endemic and naturalised species make up about 40% of Kandyan gardens (named after the city of Kandy in central Sri Lanka). Trees, shrubs, herbs, crops and animals interact and sustain themselves in association with the households. Management activities that enhance agricultural biodiversity are carried out in conjunction with domestic work or during leisure time, particularly by women. The involvement of households in managing home gardens is not limited to harvesting diverse goods. Specific management practices are frequently used to achieve better immediate use of resources, including favourable interactions. Branch pruning, coppicing, clearing excess seedlings, nurturing of naturally germinated seedlings or species emerging from garbage heaps, increasing soil nutrients, and using indigenous knowledge to control pests and experiment with genetic variations are all widespread activities. Local definitions of culture and well being cannot be considered apart from the system because of the strong link between survival of the households and agricultural biodiversity.
- In Cameroon, at least four breeds of domestic fowl (*Gallus gallus*) are kept on a free range system in villages where 70% of the national stock live. Indigenous fowl are kept for food and income generation, for gifts and sacrifices, for breeding stock and for traditional medicine preparations (Gueye, 1998). In south east Mexico, women keep as many as nine breeds of local hen, as well as local and exotic breeds of turkey, duck and broilers in their back gardens. In selecting for the best breeds, they consider eleven different characteristics,-including egg production, ease of sale, appearance, broodiness, heat and cold tolerance, growth rate and eating qualities. Using this ranking, the most preferred birds are indigenous turkeys and ducks (Intermediate Technology, 1996).

A diverse portfolio of activities based on the contributions of agricultural biodiversity (e.g. crop cultivation, harvest of wild plant species, herding, fishing, hunting) helps sustain rural livelihoods because it improves their long term resilience in the face of adverse trends or shocks. In general, increased diversity promotes more flexibility because it allows greater possibilities for substitution between opportunities that are in decline and those that are increasing. Many rural people, regardless of whether their agro-ecosystems are predominantly pastoral, swidden or based on continuous cropping deliberately incorporate wild resources into their livelihood strategies. Nor is livelihood diversification based on such wide use of agricultural biodiversity the exclusive preserve of rural households in developing countries.

In Poland for example, wild bush and berry fruits are important for local consumption and for export, with *Vaccinium myrtillus* being the principal export species at present (over 30, 000 t/year) followed by *Rubus spp., Sorbus aucuparia, Sambucus nigra, Prunus spinosa* and *Rosa spp.* (Glowacki, 1988).

Different types of agricultural biodiversity ("cultivated", "reared" or "wild") are used by different people at different times and in different places, and so contribute to livelihood strategies in a complex fashion. Understanding how cultivation, herding, fishing, collection, use and marketing of different types of agricultural biodiversity are differentiated by wealth, gender, age and ecological situation is essential to evaluate their overall economic value. Understanding this differentiation within communities is essential because there is great variation in wealth, ability, age and power in every rural society. For example, wild resources are particularly important for the food and livelihood security of the rural poor, women and children, especially in times of stress such as drought, changing land and water availability or ecological change.

These groups generally have less access to land, labour and capital and thus need to rely more on the wild diversity available. In India, the poor obtain 15-23% of their total income from common property resources, as compared with 1-3% for wealthier households (Jodha, 1986). In Zimbabwe, some poor households rely on wild fruit species as an alternative to cultivated grain for a quarter of all dry season meals (Wilson, 1990). Whilst wild food species supply vital nutritional supplements to all diets based largely on carbohydrate rich staples, they are crucial sources of vitamins and minerals for children. Children are often the most frequent collectors and consumers of wild fruit.

The degree to which the resources of agricultural biodiversity are important for local people's livelihoods affects the appropriateness of policies on resource management and on incentives for conservation and sustainable use. Comparing the economics of biological diversity use with other livelihood options can help assess people's willingness to sustain biodiversity as part of a livelihood strategy. There is however no single valid economic approach for doing this. Combining economic concepts with participatory research does nevertheless allow for a more comprehensive valuation of agricultural biodiversity, recognising not only the financial value, but also the indirect and non use values. The insights thus gained into the relative and changing seasonal importance of different types of agricultural biodiversity for livelihood security can be quite startling.

For example, "wild" agricultural biodiversity may provide a significant proportion of total household incomes, particularly where farming or herding is marginal. In parts of Botswana, where unpredictable rains make farming a risky business, basket making from the wild palm *Hyphaene petersiana*, and beer brewing from the wild fruit *Grewia bicolor* provide a more secure income source, especially for women (Bishop and Scoones, 1994). Other local level valuation studies of agricultural biodiversity conducted in a total livelihood context show that many wild resources have significant economic value by preventing the need for cash expenditure and providing ready sources of income to cash poor households, often yielding a better income than local wage labour (IIED, 1995).

The cultural and spiritual values of some parts of agricultural biodiversity can sometimes be considered as more important than monetary values. Many rural communities designate certain biological diversity-rich areas of land or water as sacred. Sacred groves, for example, are clusters of forest vegetation that are preserved for religious reasons.

They may honour a deity, provide a sanctuary for spirits, or protect a sanctified place from exploitation; some derive their sacred character from the springs of water they protect, from the medicinal and ritual properties of their plants, or from the wild animals they support. Such sacred groves are common throughout southern and south eastern Asia, Africa, the Pacific islands and Latin America. The spiritual values of sacred places on land or water are often inextricably tied with the functions that their associated agricultural biodiversity may provide in maintaining the health of the ecosystem.

For instance, in a ranking exercise conducted to show the relative importance of different values derived from savannah woodlands in Zimbabwe, villagers explained that one of the most important aspects of their woodland was the sacred areas it contained. Honouring and preserving these sacred areas according to the wishes of the ancestral spirits is essential for good rainfall. The wide range of consumption benefits derived from the woodland were ranked lower than these spiritual ecosystem functions, as they could not exist without the rains, which in turn depend on the sacredness of the woodland.

The rich tapestry of locally unique agricultural biodiversities represents, at the global level, a huge amount of diversity *between* species. Out of the 250,000 plant species that have been identified and described, some 30,000 are edible and about 7,000 have been cultivated or collected for food and the provision of other goods and services at one time or another (Wilson, 1992). World-wide, several hundred animal species including mammals, fish, reptiles, molluscs and arthropods also contribute to food and livelihood security.

Diversity *within* species is also remarkable among those plant and animal species that have been domesticated for crop and livestock production by innovative rural people. The inherent variation within farmers' crop varieties (landraces) is immense for cross-pollinated species such as millet or maize. For self-pollinated crops such as rice and barley, and for vegetatively propagated crops like potatoes and bananas, individual varieties are less variable, but the number of landraces developed may be very high. Estimates of the distinct number of varieties of Asian rice (*Oryza sativa*) range from tens of thousands to more than 100,000 while some communities in the Andes grow as many as 178 locally named potato varieties (Brush, 1991).

Livestock keepers have also generated and safeguarded considerable intra-specific diversity through their animal husbandry. In India alone, 26 different breeds of cattle and 8 breeds of buffalo, 42 breeds of sheep and 20 breeds of goat have been identified along with 8 breeds of camel, 6 breeds of horses, 17 breeds of domestic fowl,-in addition to native pigs, mithum and yak. World wide, it is believed that the total number of mammalian and avian livestock breeds is between 4,000 and 5,000. Important contributions of animal diversity to food and livelihood security are summarised in below.

Livestock Diversity's Contributions to Food and Livelihood Security

Domestic animals' contributions to livelihood security are highly site specific and seasonal, and their importance differs from one social group to another. Each contribution of livestock diversity to livelihoods is governed by many interacting institutional factors and social relations. For each economic and ecological setting, a differentiated analysis of livelihoods is therefore essential to understand what a particular contribution of livestock is worth, to whom, when and in what way. Some of the products of animal diversity include:

- *Food.* Domestic animals provide a considerable part of the food requirements of pastoralists. For small scale mixed farmers they can be especially important during the seasons of grain scarcity.
- *Nutrients.* Domestic animals are important sources of essential amino and fatty acids. Ruminants through their symbionts also convert cellulose into products that are digestible to humans.
- *Clothing.* Wool, hair and leather are used for making garments, blankets, shoes, bags etc.. These may be for subsistence or income generation.
- *Utensils.* Bone is used to make a variety of utensils, leather for making bags to carry water, food etc.
- *Construction.* Hides, wool and other fibres are used to make shelters.
- *Transport.* In some countries animals still provide the most affordable form of transport.
- *Traction.* Domestic animals reduce the amount of human labour needed for farm operations
- *Fuel.* In some places, animal dung is the only fuel available.
- *Fertiliser.* Animal manure is an important component of many mixed farming and aquaculture systems
- *Income.* Animals convert low value scattered feedstuffs into high value commodities. Since ownership is not a prerequisite for keeping animals, income derived from animal products can be especially important for the poor. Revenue from small animals may be the only income controlled by poor women.
- *Insurance.* Animals are a readily convertible source of cash and provide a safety net against unforeseen events, such as adverse climatic conditions, temporary food shortages, changes in family circumstances, and unstable commodity prices.
- *Spiritual functions.* Animals often play important roles in religious ceremonies and rituals.

Modified from: Intermediate Technology Development Group, 1996. Livestock keepers safeguarding domestic animal diversity through their animal husbandry.

Domesticated plant varieties, animal breeds and diverse agro-ecosystems are largely sustained through local peoples' crop husbandry, livestock management practices and fishing techniques. Conservation of diversity is through active use in different ecological and economic settings.

Agricultural Biodiversity's Contributions to Production and Environmental Sustainability

Each species in an agro-ecosystem is part of a web of ecological relationships connected by flows of energy and materials. Whilst each species may occupy a specific ecological niche (e.g. primary producer, specialist or generalist consumer, decomposer) it is involved in sustaining many different agro-ecosystem functions and environmental processes, either directly or indirectly. In this sense the different components of agricultural biodiversity are inherently multifunctional and contribute to the resilience of production systems whilst providing environmental services at the larger landscape level. However, it is important to note that some species may play a key driving role in forming the structure and overall behaviour of agro-ecosystems and landscapes at different scales. There is indeed growing evidence that the diversity and functional complexity of all ecosystems can be traced to a small number of critical structuring processes, some of which are mediated by critical "keystone species". An example is the suite of 35 species of insectivorous birds that mediate budworm outbreak dynamics in the eastern boreal forest of North America.

Farmers, herders and fishermen have often enhanced the multiple functions of agricultural biodiversity through choice of genetic material, design of cropping patterns, development of crop and livestock production systems, land and water management practices as well as institutional arrangements. Local knowledge about the properties and dynamic roles of agricultural biodiversity is crucial in this connection. For example, the *mal monte* (bad weeds) and *buen monte* (good weeds) management systems of Mexican farmers recognises that the vegetation community as a whole must be managed to promote those aspects that are beneficial.

In low external input farming, different components of agricultural biodiversity are usually combined to give practices finely tuned to the local biophysical and socioeconomic conditions of individual farmers, herders and fish culturists. Natural processes mediated by agricultural biodiversity are favoured over external inputs and by products or wastes from one component of the agro-ecosystem become inputs to another. An example is the mulberry grove-fishpond system in the Pearl River Delta of China. In this multifunctional system, the white mulberry (*Morus alba*) tree produces organic substances (mulberry leaves etc). These are used to feed silkworms that, in turn, produce their silk and chrysalides. The fallen parts of the mulberry tree and the excrement of the silkworm are applied to the fishpond where they are converted into fish biomass. The excrement of the fish, as well as other unused organic matter and bottom mud are returned to the mulberry grove as fertiliser, after being broken by a diverse suite of benthic micro-organisms. The agricultural biodiversity harnessed by the fish culturalists allows for the closing of nutrient cycles and efficient production in time and space. Fish polycultures are thus made up of species that dwell in the upper, medium and lower layers of the pond, as well a fish species with different feeding habits (e.g. plankton feeders, herbivorous fish, benthic mollusc feeders, and omnivorous fish).

In more specialised, high input farming based on the use of high yielding varieties, agricultural biodiversity helps sustain many production functions such as soil organic matter decomposition, pollination and pest control. In the USA or Australia for example, farmers may manage cover crops primarily to save soil and water in intensive orchard production systems. However, the species chosen will usually perform other functions in the agro-ecosystem. In addition to protecting against soil erosion, cover crops usually enhance soil structure, improve soil fertility and nutrient cycling as well as play a role in pest management by providing habitat heterogeneity and preserving a favourable balance between pests and predators. Depending on the species, trees can also provide fodder for animals, so increasing the number of internal linkages within the agro-ecosystem. These examples highlight the multiple functions of agricultural biodiversity and are intended as a reminder that functions are discussed one by one in this report only for the sake of convenience.

Decomposition and nutrient cycling functions. The crop plants, trees, livestock and fish deliberately chosen by farmers are the main determinants of the diversity of the flora and fauna that makes up the decomposer subsystem (Swift and Anderson, 1993). Available evidence shows that decomposer communities are highly diverse and are centrally involved in nutrient cycling, organic matter dynamics and other ecosystem functions. Detailed knowledge on the extent and functions of this diversity is limited; there is relatively more information on the functions of soil biodiversity than on the dynamics of decomposer communities in aquatic environments. Some functional groups in soils are widespread in distribution (e.g. nitrogen fixing bacteria, mycorrhizal fungi and predators of soil borne pests) whilst other like earthworms and termites are more restricted in their distribution. A gradually emerging picture structures soil biodiversity into a series of more or less spatially independent guilds. The spatial separation between distinct guilds (surface litter, root litter, rhizosphere guilds...) allows decomposer organisms to co-exist whilst containing communities of organisms that are functionally equivalent. For instance, several ecological guilds of earthworms can be recognised in humid tropical soils with different roles in litter transformation and as "ecosystem engineers" (Lavelle et al 1998). Through their activities of feeding, burrowing and casting, they modify the physical, chemical and biological properties of soil and thus its ability to support above ground vegetation. Together with termites, different species of earthworms are among the key functional groups in humid tropical soils. At least 42 native and exotic earthworm species common in tropical agro-ecosystems have been identified as able to resist disturbances linked to agriculture and agroforestry practices and build up sizeable populations in these environments.

The sensitivity of decomposer functions to farm management practices is also evident in the mechanised agriculture of developed countries. Comparisons of the soil biological diversity in biodynamic, organic and conventional farms in Switzerland show higher species diversity and functional levels in biodynamic and organic plots than in conventional systems. The significantly higher biomass, diversity and functional activity of soil micro-organisms, earthworms, ground beetles, staphilinids and spiders found in biological systems are largely due to the organic amendments and more selective plant protection measures used in the biological systems.

In high input-high output agriculture, microbial diversity is also a central component of integrated plant nutrition systems (IPNS) that aim to maximise the efficiency of plant nutrient supplies to crops by complementing the use of on-and off-farm sources of plant nutrients. Nitrogen fixation through bacteria (notably *Rhizobia spp*) and algae

(*Azolla spp*) as well as phosphorus cycling *via* mycorrhizal fungi species are particularly noteworthy in this connection.

Microbial diversity is generally known to mediate nutrient cycling. However, there are few detailed studies of the dynamic role of micro-organisms in structuring landscapes and agro-ecosystems at different spatial and temporal scales. A long term study of African savannahs has shown that the productivity of large mammalian herbivores,-upon which the human economy of the savannah is based, is dependent on water availability acting as an "on-off" switch for the mineralisation of nitrogen, phosphorus and sulphur. Some 80% of the mineralisation is performed by the diverse soil micro-organisms which respond dramatically to the presence of water or rain. Where the immediate limiting resource in broad leafed savannahs is nitrogen, the key effect of water is to control the availability of inorganic nitrogen by modulating the functions of soil microbial diversity rather than control photosynthesis. The interaction between water and micro-organism activity thus sustains soil fertility. In turn, soil fertility has a profound effect on savannah ecology by determining not only plant production but also what fraction of it is edible and which species of plants and animals will be present. Livestock production is thus closely dependent on the dynamic interactions between water availability and the activity of a diverse suite of soil micro-organisms in these semi-arid landscapes.

Biomass production and yield efficiency functions. Low external input production systems usually incorporate a wide range of species and genotypes that serve a variety of production goals and are used for their resistance to diseases and pests as well as for the differential exploitation of microhabitats. The relative productivity and efficiency of these diverse agro-ecosystems (fish polycultures, mixed herds, intercrops, integrated agro-silvo-pastoral systems) have been quantified in terms of relative yield or energy efficiency of diverse units as compared with sole crops. Results indicate that diversity rich agriculture is generally highly productive in terms of its use of energy and unit land area (or unit water volume). For example, the land equivalent ratio (LER) of yields with intercrops in which mixtures include a legume species is usually significantly greater than outyield of sole crops. The energy efficiency (ratio of energy output to energy input) of pig or poultry production in internally diverse agro-ecosystems can be up to 10 times higher than that of intensive pig and poultry farms based on genetically uniform single species reared with enormous subsidies of fossil fuels (Leach, 1976). Whilst the yield output per labour hour of the more intensive and uniform systems is extremely high (in the absence of the internalisation of social and environmental costs), the more agricultural biodiversity rich systems are generally efficient producers of significant amounts of biomass. This efficiency is largely a product of the systems' biological and structural complexity that increases the variety of functional linkages and synergies between different components of agricultural biodiversity.

Soil and water conservation functions. Soil, water and nutrient conservation have been improved with the use windbreaks, contour farming with appropriate border crops and cover crops in a wide range of agro-ecosystems. In France over 150 species of trees and shrubs are used for soil and moisture conservation, with different species mixtures planted as hedges and taller windbreaks in gardens, orchards, whole farms and the larger rural landscape. In Sahelian countries of Africa windbreaks made up of *Euphorbia tirucali, Parkinsonia aculeata, Opuntia tuna* and *Prosopis africana* trees and shrubs help to conserve soil and moisture, and raise the yields of cereals which are grown between. In Mexico, contour lines are often planted with *Agave americana* to conserve soils and retain moisture whilst in southern Italy and Greece *Opuntia tuna* performs similar functions. Many of these plants are multiple purpose species yielding wood, edible fruit and nuts, fodder, refuges for natural enemies of pests, nitrogen biofixation and medicines in addition to their soil and water conservation functions. Cover crops consist of plant species that are deliberately established after or intercropped with a main crop to serve various regenerative and conservation functions including soil and water conservation.

Annual bluegrass, lana vetch, crimson clover, black medic, purple vetch and barley are some of the species recommended as cover crops for orchards and vineyards in California, USA. The wide variety of management systems in high input-high output orchards and vineyards creates a demand for a diversity of cover crops. Grass species have fibrous root systems that make them particularly useful in building soil structure, providing erosion control, and improving water penetration. Legumes are not as effective as grasses in improving water penetration but they contribute nitrogen to the soil. Many cover crop options can be selected for soil and water conservation from a diversity of annually seeded winter growing grasses and legumes, summer annuals, perennial grasses and legumes and reseeding winter annual grasses and legumes.

In North America and Europe, living mulch systems can be an economic way for commercial soybean, corn and vegetable growers to reduce soil erosion and water loss, increase soil organic matter and keep yields constant in high input-high output agroecosystems. Legume species commonly used as living mulches include alfalfa, short white clover, hairy vetch and red clover.

Pest control functions. Agricultural biodiversity in the form of predators, parasitic wasps, microorganisms plays a key role in controlling agricultural pests and diseases. For example, according to CAST (1999) more than 90% of potential crop insect pests are controlled by natural enemies that live in natural and semi-natural areas adjacent to farmlands. They have estimated the substitution of pesticides for natural pest control services at a cost of $54 billion per year.

Many methods of pest control,-both traditional and modern-, rely on biological diversity. The development of crop varieties and animal breeds that are resistant to specific pests and diseases selectively draws on the genetic diversity available *in situ* and in *ex situ* collections of germplasm. Genetic mixtures deployed in temperate and tropical agroecosystems can be effective in containing diseases in small grain crops as well as insect outbreaks in cassava (Gold, 1987), corn (Power, 1988) and potato for example. There are also many documented experiences showing that insect pests tend to be less abundant and damaging in agroecosystems with higher plant diversity e.g. intercrops, polycultures, crop rotations, cover crops, mixed tree stands, mixtures of annual and perennial plants. Depending on the pest species and the context, the plant diversity acts to reduce pest damage by interfering with the host seeking and reproductive behaviour of the pest, by enhancing the pests' natural enemy populations or by a combination of these processes. Judicious vegetation management within and around agroecosystems can thus enhance biological control or confer an overall resistance to pests and disease outbreaks.

Understanding how agricultural biodiversity directly or indirectly affects pest and disease dynamics is critical for the design of pest management at different scales. For instance, recent work in Javanese rice fields shows that there is an enormous diversity of arthropods, even in high input-high output agriculture. The arthropod communities are structured such that the dynamics of seasonal succession consistently lead to high levels of pest suppression, with little chance of outbreak. From the time that water first floods a farmer's field in preparation for planting, organic matter,-derived from residues from the previous crop cycle, organic waste in irrigation water, and algal growth-, provides the energy for an array of microorganisms (bacteria and phytoplankton) and detritus-eating insects. The adults of the plankton-feeders (midges and mosquitoes) together with the detritus-feeders, provide a consistent and abundant source of alternative food for generalist predators very early in the season.

As a result, pest mortality due to predation is high from the very earliest part of the season; hence, minimising the chance of damaging pest outbreaks. However, this intrinsic strength and stability of the rice agro-ecosystem is influenced by two main, large scale, external factors: 1) local and regional patterns of pesticide use, and 2) landscape effects-specifically, the spatial scale at which fields are synchronously planted, the duration and nature of fallow periods, degree of surrounding weedy or natural vegetation and existence of nearby ponds or other sanctuaries for natural enemies. This type of information on the functions of agricultural biodiversity in rice paddies provides the ecological basis for integrated pest control through careful management of the wider landscape and decisions on pesticide use.

Pollination and dispersal functions. There are more than 100,000 known pollinators (bees, butterflies, beetles, birds, flies, and bats). Pollination mediated by components of agricultural biodiversity is an important function in a variety of terrestrial agroecosystems (biotic pollination *per se* is poorly represented in aquatic ones). About half of all plant species, including food-producing crop species, are pollinated by animals. For example, the pollination of various fruit crops by bees and other insects is critical in mountain areas of Asia.

In Nepal *Apis cerana* begins foraging at temperatures 5-7°C lower than those that initiate *Apis mellifera* foraging. Managed crop pollination with a variety of bee species in different zones plays an important role in overall agricultural development.

The benefits of pollination are also considerable in high input-high output agriculture: the economic value of pollination services in the United States is estimated in billions of dollars per year. Management practices that reduce the species or abundance of pollinators can result in less genetic variation in crops dependent on pollinator visits for reproduction, both in temperate and tropical agriculture. With a loss in pollinators, seed production declines and the vulnerability to pests and climatic change increases with the resulting loss of genetic diversity.

"Mobile link" species (i.e. animals necessary for the persistence of plant species that in turn support otherwise separate food webs) such as pollinators and seed dispersal agents may be critical to the maintenance of the species richness of tropical forest based agroecosystems and complex home gardens imitating the natural forests' architecture. Many species in tropical forests managed for food and agriculture depend on a small suite of frugivores for dispersing their seeds. Loss of these species of fruit eaters may adversely affect the long-term viability of many tree species important for food security. Reductions in genetic variability are likely to be high for plant species that are highly dependent on frugivory for seed dispersal.

Biological diversity conservation functions. There is no strict divide between "wild" and "domesticated" species important for food and livelihoods. Many wild plant species and populations that have been considered to be wild are in fact carefully nurtured by people. A similar continuum exists for animal species that use agroecosystems as habitat, nesting grounds and food.

Whilst not necessarily the subject of conscious management by herders or farmers, many wild species thrive in, or are dependent on, agroecosystems. In general the more structurally and biologically complex the agroecosystems, the more diverse the forms of wildlife. For example:

- The systems of shifting cultivation common in Asia have long been considered to be damaging to biological diversity in particular. However, given the opportunity, shifting cultivators preserve wild resources. Studies in Asia have concluded that most of the mature forests in this region are not virgin forests, but merely old forests that have reached a relatively stable dynamic state after some earlier clearing by human or natural means. Under traditional systems of shifting cultivation which speed up natural succession, wildlife flourishes, with elephants, wild cattle, deer, and wild pigs all feeding in the abandoned fields. Tigers, leopards and other predators are in turn attracted by the herbivores. Older fields contain a high proportion of fruit trees that attract primates, hornbills, squirrels and a variety of other animals.
- Long term research in the Lowland of west Poland shows that a mosaic of landscape structure of small cultivated fields, shelterbelts, meadows and small ponds helps maintain biological

diversity (and many other functions such as enhanced water storage). Mammal, amphibian, reptile, insect diversity are enhanced and the number of bird species is positively correlated with the diversity index of the landscape (Ryszkowski, 1995).

Climate functions. Although agricultural and the atmosphere interact in many ways, the links between climate and weather, and agricultural biodiversity can be rather complex.

At the global scale, the distribution of individual plants and animals, vegetation and crops is conditioned by available climatic resources such as solar radiation, which controls the production potential, and by rainfall, which determines to what extent the radiant energy can actually be used by plants for their growth. Indeed, it is for these reasons that climatic classifications largely coincide with vegetation maps.

At the local scale, types of landscapes and vegetation, including crop and crop-vegetation mixes contribute towards modifying the local climates by directly affecting wind patterns, rates of evaporation, rainfall interception (effective rainfall), etc. This in turn conditions the development of vegetation and crop canopies which can be said to create their own microclimate, resulting from the interaction of plants with the general climate. Shelter belts, or the use or large tree belts to protect tropical plantation crops from cyclones, provide clear examples in which people and communities derive direct advantage from these landscape features.

Many examples demonstrate the benefits of maintaining minimum agrobiodiversity in the face of climate variability. Recent droughts in some southern African countries for example, have shown that mixes of local varieties planted over several weeks have the potential to better resist unusual patterns of rainfall variability (e.g. early-season drought) than some modern varieties planted on the "optimum dates". In fact, there appears to be a link between crop biodiversity and relatively low but regular production, one of the keys to food security.

Similarly, atmospheric pollution has been shown to interact with agro-ecosystem functions, both positively (nutrients, such as nitrates and sulphur) and negatively (toxic compound like tropospheric ozone and heavy metals). Given that the response to these substances varies enormously across the spectrum of species, they have the potential to modify patterns of biological diversity. Agriculture also contributes towards the emission of some of the most significant "greenhouse" gases (in terms of global warming potential, GWP). Some studies have suggested for example, that increases in the area of permanently or quasi-permanently flooded rice cultivation account for a significant proportion of the increase in net methane emissions. The increased specialisation of ruminant production based on high yielding breeds has also significantly contributed to global methane emissions (methane is produced by anaerobic digestion in animals). An extreme case would be the enhanced biogenic emissions of carbonylsulphide (Hofman, 1990) associated with increasing cultivation of high sulphur crops such as rape. Oxidation of carbonylsulphide in the stratosphere leads to the production of sulphate aerosols that influence the intensity of ultra violet radiation reaching the earth's surface, potentially affecting the dynamic functions of agricultural biodiversity. Volatile organic compounds such as terpenes and isoprenes are produced and released into the atmosphere by many plant species, especially in agroecosystems dominated by Mediterranean shrubs, eucalyptus and conifers. By influencing the oxidation capacity of the troposphere, these volatile compounds influence the abundance and distribution of trace gases such as ozone. In almost all cases however, many of the these effects are marginal when compared with the climatic effects of land use change, including deforestation, which have until recently constituted one of the main non-industrial sources of carbon dioxide

Functions in the water cycle. There are few experimental studies exploring the links between agricultural biodiversity and water. However, available evidence shows that agricultural biodiversity plays a crucial role in cycling water from the soil to the atmosphere and back. It also has measurable impacts on water quality. Agroecosystems with different species assemblages and plant architectures result in differences in the amount of precipitation intercepted, the proportion of precipitation converted to stem flow, and the proportion of precipitation that infiltrates the soil rather than running off. At the landscape level, the types, relative abundances and relative spatial locations of agro-ecosystem types affect the amount of water moving from one point to another. For example, conversion of vegetation within a watershed from forest or shrub land to less structurally and biologically complex grassland is known to influence stream flow out of the watershed, in both temperate and tropical systems.

At the functional group level, the root structure, phenology and physiology of different plant species important for food and agriculture have direct implications for the quantity and timing of water transfer to the atmosphere *via* evapotranspiration. Individual plant species differ in their resistance to water stress, their efficiency of water uptake from the soil, and so on. Genetic variations among crop varieties (differences in water use efficiency, stomatal resistance...) also influence these processes. At the landscape level, evapotranspiration from agroecosystems can have an effect on relative humidity and microclimate downwind.

In both terrestrial and aquatic environments, the plant and animal components of agricultural biodiversity also function to alter water quality by performing various filtration, uptake and excretory processes. These functions all affect the composition and concentration of dissolved gases, solutes and particulates. Species level differences in physiology can positively or negatively affect water quality. However, in most cases a greater diversity of biological organisms (from microbes to fish and macrophytes) leads to a higher quality of water for human consumption and use.

The influence of agricultural biodiversity on landscape structure. A landscape is a heterogeneous area made up of a cluster of interacting ecosystems that is repeated in similar form throughout. The spatial layout between landscape elements together with the interactions and linkages between them determine the landscape's structure and functions. The many different species found in a landscape are essential components of that landscape. By providing environmental services and functions agricultural biodiversity can have a profound influence on landscape structure.

Through its positive or negative effects on agricultural biodiversity, human activity can transform whole landscapes over large areas. For example, many rural communities enrich their agricultural plots and forest fallows with valued perennial plants.

Through such enrichment practices, successional vegetation can become a site for economic production as well as for ecological rehabilitation (Dubois, 1990). Each of the major tropical forest regions has many economic woody plants that have been managed, probably for millennia, in enriched fallows (Wood, 1993). In Vanuatu the natural composition of forests has been dramatically altered by centuries of itinerant gardening, favouring tree species that bear edible fruits and nuts. Fallows have been (and still are) enriched with rattan in East Asia, rubber in Sumatra, *Casuarina* in Papua new Guinea, *Gliricidia* and peach palm in Central America, oil palm in West Africa, and edible fruits and nuts universally. Locally adapted enrichments have altered species composition and also directly influenced the structure of landscapes at different spatial scales.

The influence of agricultural biodiversity on landscape structure is partly determined by the social institutions that mediate the relationships between rural people and the environment, and partly by climate and edaphic factors. For example in the semi-arid landscapes used by pastoralists of Africa, there are high levels of spatial and temporal variability in fodder biomass production, highly variable rainfall and episodic chance events such as drought. In these non-equilibrium systems pastoralists have developed opportunistic management schemes to exploit the patchiness of the vegetation, learnt to avoid risks by moving herds and flocks to make best use of heterogeneous landscapes and diversified their livelihood activities. Pastoralists have rules and regulations which govern the use of water, pasture, animal movement and control of vegetation and trees. Management of agricultural biodiversity and the larger landscape is mediated by these local institutions.

The local adaptive management of agricultural biodiversity enables people to cope with uncertainty and sustain the dynamic environmental processes that define and shape those landscapes. This is in stark contrast with the degradation that occurs under the centrally planned, standardised rangeland and livestock management schemes often based on erroneous concepts of carrying capacity and equilibrium ecology. New perspectives in ecology have challenged the conventional views of drylands in Africa as stable ecosystems subject to decline and desertification once carrying capacity is exceeded. Rangelands and pastoral landscapes are resilient and less prone to degradation and desertification than once thought. The new findings concord with the knowledge of many local livestock herders and emphasise how rangelands are subject to high degrees of uncertainty and ecological dynamics, characterised by sudden transitions rather than slow and predictable change.

Specific components of agricultural biodiversity are often directly implicated in the processes that structure agroecosystems at different temporal and geographical scales (from small farm plots to whole water/landscapes). Even highly complex landscapes like tropical irrigated rice or forests in the savannah transition zone of West Africa, are apparently structured by a very few key variables. Research over the past 20 years in applied ecology of managed systems shows that ecosystem and landscape dynamics tend to be organised around a small number of nested cycles, each driven by a few dominant variables.

A small number of plant, animal, and abiotic processes structure biomes over scales from days and centimetres to millennia and thousands of kilometres. Individual plant and biogeochemical processes dominate at fine, fast scales; animal and abiotic processes of mesoscale disturbance dominate at intermediate scales; and geomorphological ones dominate at coarse, slow scales....the physical architecture and the speed of variables are organised into distinct clusters, each of which is controlled by one small set of structuring processes. These processes organise behaviour as a nested hierarchy of cycles of slow production and growth alternating with fast disturbance and renewal (Gunderson et al, 1995).

Identifying and understanding the dynamics of these "structuring variables" provides a practical basis for sustainable agriculture and landscape management.

Agricultural Biodiversity's Contributions to Rural Development

In addition to its direct contributions to rural livelihoods, agricultural biodiversity may generate other rural development opportunities through eco-tourism and a variety of income generating schemes. Many humanised landscapes in Europe, South America, Australia and the Asia-Pacific regions are increasingly valued for aesthetic and historical reasons. For example, throughout the Asia-Pacific region mountainous terrain has, over the centuries, been shaped into landscapes of terraced pond fields for the cultivation principally of rice, but also of taro and other crops.

In Europe, low input, extensive farming systems such as the Dehesas in the Iberian peninsula of Spain cover some two million hectares. Dehesa systems are open savannah like woodlands used as pastures, with sclerophyllous trees, mainly *Quercus rotundifolia* Lam., and a therophytic herb layer (MAB, 1989). Dehesas are home to many endangered species of wildlife such as the Iberian lynx, the golden eagle, the little bustard and the Egyptian vulture.

These landscapes exist both as archaeological (i.e. preserved) sites and as living landscapes, which continue to be used and maintained by the people who created them. The conservation of these cultural landscapes is considered important by a growing number of stakeholders. For example, many low external input agroecosystems in Europe are valued by urban populations ready to pay for the experience of a holiday in rural areas (IEEP and WWF, 1994). At a global level, the intrinsic value of THESE cultural landscapes, and what they can teach about enduring systems of human-nature interaction, has led to a strategy within which the identification, evaluation and conservation of specific regional landscape types are to be considered within the framework of the World Heritage Convention (UNESCO, 1996). The ecotourism potential of these cultural landscapes is viewed as potentially important for rural development and local employment creation, both in the developed and developing countries. However, recent evidence suggests that the potential of eco-tourism can only be realised under certain conditions.

As is often the case for classical tourism, eco-tourism schemes tend not to be integrated with other sectors of the national or regional economy; and only a fraction of earnings generated actually reach or remain in the rural areas. More importantly, the majority of the rural population is frequently bypassed economically even where some earnings remain in the tourist location, as they are used up by the related administration or appropriated by local elites and business people. At the same time, local livelihood sources and cultures are negatively affected in nearly all cases. Generating economic benefits and fostering equitable rural development is only feasible when many wide-ranging reforms,-such as restoration of land and water rights to local communities, support for new forms of tenure and rights of usufruct, strengthening of local groups and institutions, investment in technical and managerial skills and mandatory impact assessments of all ecotourism schemes-, are carried out (Koch, 1997). The necessary structural political and economic changes along these lines are difficult in many developing and developed countries.

Another potential engine for rural development is the exploration, extraction and screening of biological diversity and indigenous knowledge for commercially valuable genetic and biochemical resources (biodiversity prospecting or bioprospecting). A detailed treatment of the economics of bioprospecting is beyond the scope of this paper. However, it may be noted here that despite frequent mention of benefit sharing agreements in commercial contracts between bioprospecting agents and sovereign states, the specific terms of benefit sharing are strictly confidential. Available evidence indicates that benefits shared with countries in which collections took place represent a small fraction of the annual Research and Development budget of the corporations involved. Moreover, indigenous and local people receive only a minuscule proportion of the profits generated from sales of products that embody their knowledge and resources. For example, one study has estimated that less than 0.001 per cent of the market value of plant based medicines has been returned to local and indigenous peoples from whom much of the original knowledge came. And while various codes of conduct and guidelines have been developed to ensure greater equity, compensation and fair sharing of benefits between bioprospecting companies and local communities, none are internationally legally binding.

Further opportunities for rural development hinge on creating local businesses and products that sustainably use agricultural biodiversity and add value to it in the context of more localised economies. In a growing number of rural areas in Europe, North America, Australia, Japan and New Zealand the diversity of local plants and animals is being harnessed for sustainable economic development. In south east France, the regional genetic heritage program of the Provence-Alpes-Cotes d'Azur involves a wide range of actors spread across six administrative departments,-about 31,500 square kilometres of very diverse ecosystems. Ways and means of reintegrating locally adapted, traditional animal breeds (sheep, goats, cattle and bees) and crop varieties (fruit trees, fodder plants and cereals) are being explored to generate local products, jobs, income and environmental care. Similar initiatives are reinvigorating local economies and employment in the Willapa watershed of the Pacific North West (USA). The Willapa watershed includes 275,000 hectares on the coast of Washington state and is rich in agricultural biodiversity: oysters, clams, sturgeon, crabs, salmon and dense forests. With the support of the Ford Foundation and a Chicago based community bank a range of local businesses have been set up to add value to this local agricultural biodiversity. For example, Willapa oysters are now marketed locally rather than shipped out wholesale, alder is harvested from secondary forests for high quality wood products, fish and crab are marketed with the northwest image of wholesome foods, cranberry growers produce a wide range of products retaining more of the value added from food processing within the watershed.

These forms of endogenous rural development seek to create viable and locally controlled economic activities based on locally adapted agricultural biodiversity, knowledge, skills and negotiated partnerships between civil society, government and the private sector. Initiatives to reclaim diversity for rural development often focus on regenerating local food systems and economies based on comprehensive definitions of well being and wealth, both in developed and developing countries.

These examples together with the recent "discovery" of the potential of cultural landscapes and the creativity of their inhabitants illustrate a more fundamental point. Agricultural biodiversity, together with the local knowledge, institutions and management practices associated with it, may provide a robust foundation for development in many different economic and ecological settings. Understanding the forces that have neglected or undermined the values and functions of agricultural biodiversity can help identify ways forward.

Underlying Causes of Agricultural Biodiversity Losses

The Neglect of Indigenous Knowledge, Local Institutions and Management Systems

Local and indigenous ways of knowing, valuing and organising the world have been frequently brushed aside by so called "modern" technical knowledge. Both colonial and post-colonial administrations in Africa, Asia and the Americas, as well as governments in the newly emerging nation states of Europe, have often either neglected or undermined local management systems. These systems were generally sensitive to the needs of local people and often enhanced their capacity to adapt to dynamic social and ecological circumstances. In addition to ignoring local knowledge and skills, many modernising interventions have superseded or substituted existing formal and informal institutions that were central for the sustainable management of agricultural biodiversity.

In both terrestrial and aquatic landscapes, these local groups enforce rules, incentives and penalties for eliciting behaviour conducive to rational and effective resource conservation and use. For as long as people have engaged in livelihoods pursuits, they have worked together on resource management, labour sharing, marketing and many other activities that would be too costly, or impossible, if done alone. Local groups and indigenous institutions have always been important in facilitating collective action and coordinated management of agricultural biodiversity at different spatial scales. Indigenous peoples resource management institutions provide striking examples of active conservation and sustainable use. These institutions include rules about use of biological resources and acceptable distribution of benefits, definitions of rights and responsibilities, means by which tenure is determined, conflict resolution mechanisms and methods of enforcing rules, cultural sanctions and beliefs (Alcorn, 1994).

This neglect of human ingenuity and diversity continues today and ultimately reinforces the dominant model of development based on uniformity, centralisation and control (Scott, 1998).

The Dominance of Blueprint Paradigms and Policies

One of the most fundamental causes of agricultural biodiversity loss, both past and present, is the active promotion and spread of the blueprint approach to development. Typical expressions of this are industrial agriculture and the closely related Green Revolution. The blueprint paradigm is also reflected in many contemporary forest, fishery and rangeland development. This approach to food and fibre production focuses on maximising commercially important yields and productivity through the use of monoculture systems and uniform technologies, including high yielding seeds and animal breeds, agrochemicals, irrigation, mechanised equipment and large infrastructure developments. The methods and means deployed for production largely originated in the West where money and trained personnel ensure that technologies work and that laws are enforced to secure management objectives. During and after the colonial period, these technologies, and the values associated with them, were often "exported" from the North to the South.

Managerially, the blueprint approach fits the type of organisations with clear and fixed definitions of roles, procedures and methods, hierarchical authority, punitive management style and inhibited lateral communications. These organisational structures are often better suited to routine activities and do not cope well with fast changing circumstances. The main actors in these organisations are normal professionals who are concerned not just with research, but also with action. Normal professionals are found in research institutes and universities as well as in international and national organisations where most of them work in specialised departments of government (forestry, fisheries, agriculture, health, wildlife conservation, pastoral administration...). Despite some notable exceptions, the thinking, values, methods and behaviour dominant in their profession or discipline tends to be stable and conservative. Lastly, normal professionalism generally "values and rewards "first" biases which are urban, industrial, high technology, male, quantifying, and concerned with things and with the needs and interests of the rich". This blueprint approach to the management of agricultural biodiversity is often supported, subsidised and defended by an elaborate institutional structure,-including many international donors and development agencies, international and national research institutions and national governments. Numerous policies,-ranging from general agricultural development policies to pricing and credit schemes-, directly or indirectly influence biological diversity in livestock production, agriculture, forestry and fisheries.

The most influential are incentive policies (e.g. subsidies for agrochemical inputs, extension programs, credit policies and marketing standards) that support the adoption of capital and energy intensive industrial inputs and technologies. For example, extension programs in many countries have mandated the adoption of uniform varieties and the elimination of diversity. Policy incentives for people to clear forested land and establish farms in order to gain tenure in Brazil, Costa Rica and Indonesia have contributed to increases in food production but they have also induced biological diversity losses and unsustainable land use.

Corporate Interests

Private companies, particularly transnational corporations that market agricultural inputs and process food and fibres, exert a strong influence on the type of agricultural biodiversity used in production. During the 1940s and 1950s, research and development (R&D) capabilities started to move out of public institutions into the hands of the private sector.

By the late 1990's, the pace of corporate concentration in the food, agrochemicals, pharmaceuticals, seeds and animal veterinary products accelerated. In 1998, the top ten seed companies controlled approximately 32% of the US $23 billion seed trade world-wide whilst the top ten animal health firms control about 60% of the US $17 billion animal health industry. Over 85% of the US$ 31 billion agrochemical market is controlled by less than eight corporations. With the help of new biotechnologies (e.g. gene splicing, enzyme technology) traditional boundaries between pharmaceuticals, agribusiness, biotechnology, food, chemicals, cosmetics and the energy sector are becoming increasingly blurred. Biology and the use of the diversity offered by plant, animal and microbial genetic resources is the common denominator. In many countries, including in the OECD countries, the R&D budget of these corporations dwarfs that of public sector research. As a result, corporate priorities and industrial strategies are increasingly reflected in research, development and distribution of seeds, livestock and other technologies that directly affect agricultural biodiversity. Evidence suggests that the corporate quest for commercial profits and control over production has promoted more, rather than less, genetic and ecological uniformity in agroecosystems. For example,, new biotechnologies such as pesticide resistant crops and seeds engineered to terminate germination after one growing season constitute potentially serious threats for agricultural biodiversity, at different temporal and spatial scales.

Concerns are increasingly voiced regarding the level of influence some corporations have in determining which areas of scientific knowledge are developed, and for whom. The private sector invariably privileges planning and investment directed at short-term returns rather than longer-term ones. Over time for example, a range of more reductionist scientific perspectives and techniques have been selectively favoured over whole ecosystem science approaches, basic taxonomic work, population biology, landscape ecology and understandings of human-environment interactions based on plurial and interdisciplinary perspectives. This seriously undermines the long term ability of society to design sustainable agroecosystems based on a functional agricultural biodiversity that reduces dependence on suppliers of off farm inputs. Moreover, market dominance combined with monopoly patents gives the life industry unprecedented control over the products and processes of agricultural biodiversity,-the biological basis of food and livelihood security.

Inequitable Tenure and Control over Resources

A significant cause of agricultural biodiversity loss is linked with the inequitable access to, and control over, land, water, trees and genetic resources. Denying rights of access and resource use to local people severely reduces their incentive to conserve resources and undermines local livelihood security. Both colonial and many national governments have a long history of denying the rights of indigenous peoples and rural communities over their ancestral lands and the resources contained therein. Denial of access, insecure tenure and rights of usufruct over the agricultural biodiversity contained in protected areas is one of the major factors undermining both conservation and development objectives. The same is true for forests, wetlands, farms, rangelands and common property lands outside of protected area networks. Recognition of anthropogenic landscapes and "wild" species moulded by human agency has important implications for ownership, and consequently rights over access and use of biological resources. However, Western concepts of private property do not recognise the intellectual contributions and informal innovations of indigenous and rural peoples who have modified, conserved and managed so called "wild" species and landscapes (Crucible, 1994).

Inequities in access and control over genetic resources of domesticated plant and animals have also contributed to the erosion of diversity and the exploitation or displacement of local knowledge. Although most genetic resources originate from developing countries, transnational companies and northern institutions have captured a larger share of the benefits from using such resources in breeding programs and new natural product development. Legal means such as industrial patents and other intellectual property rights allow companies and northern institutions to maintain disproportionate control over the knowledge, genetic resources and benefits associated with agricultural biodiversity. In contrast, the local communities and farmers who originally nurtured this genetic diversity have generally not been recognised nor compensated for their innovations.

Market Pressures and the Undervaluation of Agricultural Biodiversity

Even though agricultural biodiversity has many values and performs many functions, it is undervalued or even ignored in conventional economic assessments. This is partly because the multiple ecological functions of agricultural biodiversity are difficult to value in economic terms. Moreover, the few economic analyses of biological diversity conducted so far have essentially focused on global values and foreign exchange elements and very little on the household use values of, for example," wild" foods and medicines. Simple economic valuations based on direct use values (for consumption or sale) have often been misleading and too reductionist to provide a sound decision making basis for policy makers and land use planners. The economic and social values of much of the biological diversity that nurtures rural people have been ignored or underperceived by outside professionals. This has biased conventional resource planning in favour of major food crops and species of commercial importance for urban centres.

The expansion of global markets and recent patterns of trade liberalisation tend to have a homogenising effect on agricultural biodiversity by standardising food production and consumption. Global markets usually demand uniform foods that are increasingly processed and sold by transnational corporations, and are geared to meet the food desires of relatively wealthy, urban based consumers-both in developing and developed countries.

In turn, these market pressures often force farmers worldwide to comply with those demands for uniformity. The policies for harmonisation of standards that accompany the globalization of markets are also powerful forces undermining efforts for the sustainable use of local agricultural biodiversity and local adaptations.

Demographic Factors

The expansion of human populations and large migrations are often partly responsible for agricultural biodiversity losses in new "frontier" areas such as forests, coastal zones, mangroves and grasslands. Whilst in some contexts population growth *per se* is clearly responsible for agricultural biodiversity loss, there are many situations in which inequitable land tenure, forest concession policies, colonisation programs, land use and fishing policies are the root causes behind the biological diversity loss induced by growth in human numbers or migrations.

Conversely, more people can mean more care for the environment and enhanced agricultural biodiversity under certain conditions, as shown by research in Sierra Leone (Richards, 1993) and Kenya (Tiffen et al, 1994).

Options for Sustaining Agricultural Biodiversity and its multiple functions

A recent FAO-SCBD International Technical Workshop, held in December 1998, proposed a series of actions necessary for sustaining agricultural biodiversity and agro-ecosystem functions. This concluding section endorses and complements the proposed actions, and presents additional recommendations which expand on the results of the workshop and flow on from the analysis of the root causes of agricultural biodiversity loss outline. Taken together, these recommendations and action points offer a contribution to the formulation of a comprehensive policy framework for national governments and international organisations.

Actions proposed by the joint FAO and SCBD International Technical Workshop on Sustaining Agricultural Biodiversity and Agro-Ecosystem Functions, 2-4 December 1998, Rome. The workshop concluded that the following four sets of actions for the conservation and sustainable use of all agricultural biodiversity, especially at agro-ecosystem levels, should be prioritised, bearing in mind that many of these actions have already been identified for particular sectors or types of agricultural biodiversity by other forums.

Information, Assessment and Indicators

Despite the work of many organisations on the development of assessment methodologies and indicators, the workshop identified deficiencies with respect to agricultural biodiversity at agro-ecosystem levels and prioritised the following needs:

- to facilitate the exchange of information between different actors and stakeholders;
- to identify agro-ecosystem-specific indicators and the use of these for assessment, monitoring and understanding the causes of changes in agricultural biodiversity;
- to focus on developing indicators particularly for changes at agro-ecosystem levels and for the economic forces that influence these changes;
- to link indicators and assessment with particular dimensions of agricultural biodiversity, such as for food security, biological support systems or agro-ecosystem functions.

Research and Development

Although the research and development programmes of many international, national and local organisations already have focused on activities for the conservation and sustainable use of agricultural biodiversity, the workshop prioritised the need for:

- emphasising greater coordination and information sharing between research and development programmes and better formal and informal sector linkages;
- strengthening national agricultural research systems on agricultural biodiversity related issues and for increasing research to demonstrate the value and costs and benefits of agricultural biodiversity conservation and sustainable use, building on proven local practices wherever possible;

- furthering farmer-driven participatory research and technology development processes through farmer field schools, recognition of local knowledge systems, etc., with full participation of local communities;
- emphasising three main issues considered essential for research and development: ecosystem approaches and ecosystem functions; specific research on classes of species such as soil biota, pollinators and predators that are essential for productive soils and plants; and threats and positive incentives for agricultural biodiversity.
- developing a set of guiding principles for the identification, development, evaluation and reproduction of ecologically sound production systems and agricultural practices which promote the conservation and sustainable use of biodiversity in agricultural ecosystems;
- promoting communications and facilitating the exchange of information on relevant scientific, research and practical information among different actors and stakeholders, including practitioners, producers and their organisations, and to relevant institutions and decision-makers, especially through improved South-South exchanges of electronic information.

Awareness Raising and Capacity Building

Despite the interventions and actions of FAO, CBD and many expert institutions at all levels, and increased attention to biological diversity and sustainable use issues since UNCED, and bearing in mind the ecosystem approach adopted by the COP, the workshop prioritised actions for:

- capacity building to improve knowledge and information on agricultural biodiversity-which remains a key issue that hinders greater commitment and support-especially by raising awareness of the value and importance of agricultural biodiversity at the agro-ecosystem level;
- capacity building to disseminate sustainable methods for agricultural biodiversity conservation by demonstrating through case studies, training and briefing materials and public media as well as field demonstrations the importance and value of agricultural biodiversity in diverse agro-ecosystems and landscapes for all types of production systems;

- capacity building for decision-making and planning and policy-making on agricultural biodiversity by increasing communication, training and information campaigns in order to raise awareness and dialogue among policy-makers, politicians, professionals, producers, consumers, the public and students.

Development of Policies and Instruments

Even though there are a number of separate decisions, instruments, policies and programmes that address aspects of the conservation and sustainable use of agricultural biodiversity in agro-ecosystems, the workshop prioritised the need for:

- integrating agricultural biodiversity in national biodiversity programmes and action plans as well as in national environmental action plans and agricultural strategies and plans;
- developing coordination and policy coherence at international, national and regional level between relevant organisations, ministries and sectoral bodies at all levels;
- mitigating the influences of (and reforming where possible) the market, market forces and the existing economic framework which have major impacts on agricultural biodiversity, exacerbated by economic disincentives through, for example, inequitable land tenure and negative or perverse incentives;
- introducing incentive measures as important instruments to counter the above, including: fees, charges and environmental taxes; certification and eco-labelling; market creation and property rights; and regulations;
- developing and implementing a Code of Conduct on Agricultural Biodiversity, based on existing agreements, which would assist private sector, government and civil society organisations to identify their rights and obligations and inform their policy-makers and programme developers.

Broadly speaking, there are two alternative scenarios for the management of agricultural biodiversity. The dominant blueprint approach to development has been identified as a major underlying cause of agricultural biodiversity loss. Nevertheless, national governments, the private sector and civil society may choose to stay within this paradigm and reform some of its less acceptable elements in their quest for more sustainable agriculture and land use.

In sharp contrast, the learning process approach focuses on reversals from the norm and structural change, rather than systemic adjustments within well defined and often narrow boundary conditions. Discussions around these alternative scenarios are inevitably emotionally charged. The issues at stake go beyond purely technical matters and include the fundamental human right to food, the right to a healthy environment as well as the political economy of who gains and who loses.

These are difficult political questions that require contradictory debate within society and negotiated solutions involving all stakeholders. Answers are not the prerogative of experts and technical bodies alone. All the latter can do is to facilitate public debate by highlighting possible policy options and technical choices. Whilst some policy recommendations presented below may be relevant for both scenarios, many have been framed within the scenario that departs from dominant values and practices.

Expand Knowledge on the Dynamics of Agricultural Biodiversity

Rationale

Knowledge about agricultural biodiversity and the processes enhancing or eroding it is the basis of national and international policy making and determines which kinds of management regimes prevail. However, much is uncertain and unknown about the structure and multiple functions of agricultural biodiversity.

There are huge gaps in knowledge on the number of species living on Earth: estimates of total species numbers vary between 5 and 30 million and a mere 1.6 million species have been described to date. Knowledge on the functions of biodiversity,-synergies and complementarities, interactions within agro-ecosystems, ecological processes within soils and interactions with the atmosphere and water, are rudimentary.

An emerging picture describes the structure and functions of agricultural biodiversity in terms of variability, sudden as well as slow change, complexity and indeterminancy at different spatial and temporal scales. But there are considerable uncertainties and on-going scientific debates on the actual functioning and dynamics of ecosystems and landscapes (e.g. equilibrium versus disequilibrium ecology, views on succession, stability-diversity relationships, carrying capacity...). Major investments are therefore needed to improve and expand our knowledge about agricultural biodiversity and its functions. Historical analysis, the use of complementary methods from the social and natural sciences and the knowledge of local resource users are all clearly needed to identify and properly explain the structure and functions of agricultural biodiversity at different scales.

There are, after all, differently situated forms of knowledge about agricultural biodiversity, and each is partial and incomplete. Participatory learning and action is needed to bring together these multiple and separate realities, combining the strengths of modern science with local knowledge.

There is indeed a strong rationale for democratising science in an age of uncertainty by directly involving "extended peer communities" (Funtowicz and Ravetz, 1993) that include farmers, herders, forest dwellers, fisherfolk and other rural people in the production and sharing of knowledge on agricultural biodiversity and its many functions.

Actions

- Provide adequate fiscal and administrative support for basic taxonomic work and inventories within and among plant, animal, microbial species and varieties
- Provide support for studies aimed at understanding the dynamic functions of agricultural biodiversity at different spatial and temporal scales, with a particular focus on the roles of soil biodiversity, pollinators, pest predators and the processes of landscape transformation
- Develop and use methods and indicators to monitor the impacts of agricultural extensification and intensification on biological diversity and local livelihoods, and promote their application
- Provide support and high rewards for studies on the functions of agricultural biodiversity that combine indigenous with scientific knowledge, using innovative participatory and complementary methodologies to do basic research in an applied way
- Diversify the governance and the membership of budget allocation committees of public sector planning and research institutes to include representatives of farmer, pastoralist, forest and fishing communities, organisations and federations, at both local and national levels

- Establish procedures to ensure transparency, equity and accountability in the allocation of research funds and dissemination of new knowledge.

Increase the Effective use of Agricultural Biodiversity in Food and Fibre Production

Rationale

Agricultural biodiversity performs vital functions in agriculture, land and water use. The diversity of plants, animals and microorganisms is essential for maintaining the productivity and sustainability of farm crops and animals, managed forests and rangelands, aquaculture and fisheries. Future global food security is dependent on harnessing and sustaining agricultural biodiversity and its many functions, from the farm plot to the landscape level.

In both low external input and high input agriculture, the goals of sustainability, productivity and equity may best be met through agro-ecosystem designs that enhance functional diversity at the genetic, species and landscape levels. A central challenge across the whole range of agroecosystems is to find alternatives to the input substitution approach and future dependence on costly biotechnology packages.

This can be achieved through an agro-ecological approach that seeks to break the monoculture structure and dependence on suppliers of off farm inputs through the design of integrated agroecosystems.

By assembling a functional biological diversity within and around agroecosystems, it is possible to encourage synergism's that subsidise agro-ecosystem processes by providing ecological services, the recycling of nutrients and the enhancement of natural enemies of pests as well as provide diverse, quality foods and other farm products.

The current emphasis on genetic engineering must be balanced by higher level approaches that build on agro-ecology, landscape ecology as well as social and biological diversity. National sovereignty and food security ultimately depend on a wide choice of agricultural technologies and development options.

Actions

- Establish and implement national policies to support agro-ecological approaches that enhance agricultural biodiversity in food and fibre production through integrated pest, crop, nutrient and soil management as well as land use planning. The policy framework should ensure political commitment, educational and institutional capacities as well as incentives for enhancing agricultural biodiversity and its functions in agriculture and land use

- Provide fiscal and administrative support for studies that investigate ways to maintain and enhance the utilisation of agricultural biodiversity in crop and animal production and in different kinds of agroecosystems, particularly high yielding and intensive commercial production systems
- Promote development on the basis of locally adapted genetic material. Strengthen in particular the capacity to develop new crop varieties and animal breeds that are specifically adapted to local environments. Increase the range of genetic diversity available to farmers
- Broaden the use of genetic diversity to protect crops and animal breeds against pest and weather problems by introducing multiple genetic systems for coping with stresses and also by deploying functional genetic mixtures and multilines where appropriate
- Broaden the use of species diversity into functional designs for agroecosystems that sponsor more of their own soil fertility, crop protection, pollination and water management (agroforestry systems, multiple cropping, fish polycultures...).
- Plan and manage rural landscapes to sustain biological diversity and ecosystem services. Where appropriate, maintain hedgerows, windbreaks and mangrove strips, leave tracts of land in native habitat, plant a diversity of crops, encourage pastoral activities, mixed species forestry and aquaculture. Maximise the use of resources internal to the landscape whilst closing nutrient cycles by integrating production with local needs and markets situated within or near the managed landscape. Apply the same principles for urban landscapes where food and fibre are produced.

Promote Local Adaptive Management of Agricultural Biodiversity

Rationale

Variation within and among agroecosystems is enormous. Daily, seasonal and longer term changes in the spatial structure of agricultural biodiversity are apparent at the broad landscape level right down to small plots of cultivated land. These spatio-temporal dynamics have major implications for the way agricultural biodiversity is managed,-how, by whom and for what purpose.

Uncertainty, spatial variability and complex non-equilibrium and non linear ecological dynamics emphasise the need for flexible responses, mobility and local level adaptive resource management in which local users of agricultural biodiversity are central actors in analysis, planning, negotiations and action.

This calls for far greater appreciation of local farming practices and knowledge used by rural people to manage agricultural biodiversity in forests, wetlands, fields, rangelands, coastal zones and freshwater systems.

It frequently suggest new practical avenues for technical support in which land users' own priorities, knowledge, perspectives, institutions, practices and indicators gain validity.

Spatial and Temporal Variation in Agricultural Biodiversity: Some Management Implications

- Crop varieties planted out experience rapid changes in environmental conditions, both above and below ground. For example, the physico-chemical and biological characteristics of soils are rarely homogenous within a single plot, let alone between plots. The intense selective pressures associated with this kind of microgeographical variation calls for a fine grained approach to agricultural biodiversity management that hinges on participatory plant breeding and decentralised seed multiplication. This adaptive strategy is generally advocated for resource poor farming systems in marginal, risk prone environments. However, it may be increasingly relevant for high input situations where agricultural diversification is used to solve production problems induced by genetic uniformity (e.g. pest outbreaks) or to exploit new market opportunities (economic niche for local or regional products).
- The abundance of insect pests and their predators is known to vary enormously within and between fields, even in the more intensively managed systems. In high input irrigated rice farms, 100 fold differences in abundance of planthopper populations are commonly observed on rice plants grown a few metres apart. Huge variations in insect abundance also exist

at larger spatial scales and all are marked by dynamic change over time. This implies that highly differentiated pest management approaches are needed to monitor and control pests effectively and economically.

The FAO-Government program on IPM is a clear demonstration of the advantages of such local adaptive management of pests and their predators in irrigated rice in Asia.

Farmer Field Schools have been a major innovation for the local adaptive management of agricultural biodiversity by developing the capacity of farmers to develop site-specific crop protection solutions.

Actions

- Carry out administrative tasks, land use planning, agricultural research and development as near to the level of actual users of resources or beneficiaries of administration as is compatible with efficiency and accountability
- Ensure flexibility and diversity in institutional and organisational design to enable government administration and services to track appropriately the dynamic changes which occur in agroecosystems and the functions of agricultural biodiversity at different time and spatial scales.
- Educate policy makers, professionals and the public (including the bearers of local knowledge) about the value of local and indigenous knowledge and management systems in sustaining agricultural biodiversity and its many functions
- Strengthen local groups and institutions by devolving resources and removing administrative or legal hurdles to local planning and action. Support the development of local institutions for common property resources and the equitable sharing of benefits from their utilisation
- Identify and support a mediator for conflict resolution and an arbiter of last resort; guaranteeing a level legal playing field and equality of advocacy in disputes, both within and between local groups as well as between local groups and powerful external interests. Of particular importance in this connection are government policies and actions that explicitly prevent discrimination on the basis of differences in gender, ethnic origin and wealth.

Support Local Participation in Planning, Management and Evaluation

Rationale

From the outset, the definition of *what* agricultural biodiversity is to be conserved, *how* it should be managed and *for whom* should be based on interactive dialogue to understand how local livelihoods are constructed and people's own definitions of well being. Most professionals have tended to project their own categories and priorities onto local people and landscape management. In particular, their views of the realities of the poor, and what should be done, have generally been constructed from a distance. Household livelihood strategies often involve different members in diverse activities and sources of support at different times of the year. Many of these, like collecting wild foods and medicine, home gardening, common property resources, share-rearing livestock and stinting are largely unseen by outside professionals.

Agricultural Research and Development and land use planning should start with enabling local people, especially the poor, to conduct their own analysis and define their own priorities. This methodological orientation is absolutely essential to meet the goals of equity, sustainability, productivity and accountability. In that context, dialogue, negotiation, bargaining, conflict resolution and joint management agreements are all integral parts of a long term participatory process which continues well after the initial appraisal and planning phases into monitoring and evaluation. This implies the adoption of a learning process approach in the management of agricultural biodiversity and its functions. It also calls for a new professionalism with new concepts, values, participatory methodologies and behaviour (Pretty and Chambers, 1993).

Actions

- Ensure public participation of women and men (particularly farmer, herder, fisherfolk and forest dweller involvement) in the development of land use and agricultural policies and in the generation of technologies
- Ensure inclusive equitable representation (gender, class, ethnic origin, age) in the participatory activities and process
- Provide capacity building for technical and scientific personnel to foster those participatory skills, attitudes and behaviour

needed to learn from farmers and rural people (mutual listening, respect, gender sensitivity as well as methods for participatory learning and action)

- Provide institutional space and incentives for professionals to understand social and cultural complexity as well as agro-ecological diversity
- Support joint problem-solving among local people, scientists and/or extension workers and the development of negotiated participatory research agendas and resource management agreements, using local criteria and indicators as well as those of outside professionals and their organisations
- Support the participatory monitoring and evaluation of national policies, land use plans, food and fibre production technologies by building on the perspectives of all social actors. Encourage the use of gender desegregated and socially differentiated local indicators and criteria in monitoring and evaluation as well as in guiding subsequent technical support, policy changes and allocation of scarce resources for agricultural biodiversity management.

Transform Bureaucracies and Professional Practice

Rationale

Local adaptive management of agricultural biodiversity and large-scale participation do not mean that state bureaucracies and other external organisations have no role. But they do challenge bureaucracies to assume different roles and responsibilities. In particular, existing bureaucracies and professionals will often need to shift from being project implementers to new roles that facilitate local people's analysis, planning, action, monitoring and evaluation.

The whole process should lead to local institution building or strengthening, so enhancing the capacity of people to take action on their own. Appropriate partnerships and co-management agreements between states, the private sector and rural communities are also required through new legislation, policies, institutional linkages and processes. However, the adoption of a participatory culture and changes in professional attitudes and behaviour are unlikely to automatically follow when new methods are adopted. Training of agency personnel in participatory principles, concepts and methods must be viewed as part of a larger process of reorienting institutional policies, organisational cultures, procedures, financial management practices, reporting systems, supervisory methods, reward systems and norms. In both government departments and non governmental organisations, the challenge for top and middle management is to design appropriate institutional mechanisms and rewards to encourage the spread of participatory methods within the organisation.

Without this support from the top, it is unlikely that participatory approaches that enhance local capacities and innovation will become core professional activities.

An important challenge concerns the creation of new relationships between local communities and their representative organisations, government agencies, research institutions and transnational corporations. Capitalising on the energy, resource efficiency and dynamism of the private sector constitutes a significant opportunity for increasing the performance of the agricultural sector and environmental sustainability.

Actions

- Encourage shifts from hierarchical and rigidly bureaucratic structures to "flat", flexible and responsive organisations
- Ensure that senior and middle management positions are occupied by competent facilitators of organisational change, with the vision, commitment and ability to reverse gender and other discriminatory biases in the ideologies, disciplines and practices animating an organisation.
- Promote and reward management that is consultative and participatory rather than verticalist and efficiency led. Establish incentive and accountability systems that are equitable for women and men
- Provide incentives and high rewards for staff to experiment, take initiatives and acknowledge errors as a way of learning by doing and engaging with the diverse local realities of farming, fishing and pastoral societies
- Redesign practical arrangements, the use of space and time within the workplace to meet the diverse needs of women, men and older staff as well as their new professional obligations to work more closely with farmers and other stakeholders (time tables, career paths, working hours, provision of paternity and maternity leave, childcare provisions, mini sabbaticals, promotion criteria...)

- Support the creation of mechanisms of communication and collaboration between the private sector, and in particular, major corporations active in the agricultural sector, and other kinds of institutions and groups

Strengthen Local Rights and Security of Tenure

Rationale

There is a need to provide a legal framework within which a devolved management of agricultural biodiversity can operate effectively, especially in respect of resource tenure. The legitimacy of rural peoples' claims to tenure and rights to agricultural biodiversity are made more apparent as landscapes are re-interpreted as the product of social and ecological histories.

If landscapes, species and genetically distinct populations have been moulded or modified by human presence and actions, local communities may claim special rights of access, decision, control and property over them.

These findings support a rights based approach to the participatory management of biological diversity important for food, agriculture and livelihoods. They also have major implications for national policies on the sharing of benefits derived from the use of landscapes, agricultural biodiversity and its end products. Guaranteeing the secular right of farmers' to save and re-use seeds and livestock progeny is crucial in this connection.

Failure to enshrine these rights in national legislation and policy practice may lead to inequitable benefit sharing schemes and conflicts that could undermine the sustainable management of agricultural biodiversity and food security.

Actions

- Reform policies and laws on rights of access, use and control over land, trees, water and genetic resources to ensure that farmer's and indigenous peoples' rights are protected as a basis for equitable benefit sharing arrangements
- Ensure that intellectual property rights over genetic resources currently re-negotiated within the TRIPs agreement of the World Trade Organisation do not undermine the objectives of conservation and sustainable use mandated by the Convention on Biological Diversity and ongoing negotiations on the International Undertaking on Plant Genetic Resources

Reform Trade-related Policies, Markets and Economic Incentives

Rationale

Economic instruments are key to sustaining agricultural biodiversity and its multiple functions.

Trade policies, markets, subsidies and economic incentives must reinforce the objectives of the Convention on Biological Diversity rather than contradict or actively undermine them.

A multilevel and systemic approach to economic transformation will often be needed to reform trade, taxation and public spending aimed at sustaining agricultural biodiversity and its multiple functions.

Actions

- Reform international and national trade-related policies that contribute to the loss of agricultural biodiversity and develop trade rules that promote social and biological diversity by regenerating local economies and food systems
- Eliminate policies and economic incentives that contribute to the erosion of agricultural biodiversity and its functions, particularly subsidies for High Yielding Varieties, pesticides and fertilisers; credit policies that require the use of such inputs and monocultures; variety release and seed certification legislation that hinder the utilisation of diverse genetic material through their requirements for distinctiveness, uniformity and stability; pricing and tax policies that favour genetically and ecologically uniform production systems
- Assess the economic benefits of agricultural biodiversity (domesticated and managed wild diversity) in a more comprehensive manner to improve the decision making basis for policy makers, land use planners and agricultural R&D.

  Economic benefits based on the use of « wild » and domesticated agricultural biodiversity should be situated and evaluated in a total livelihood context
- Establish flexibility in marketing standards to allow food distributors and retailers to diversify varieties of produce and reduce wasteful cosmetic standards for foods in markets

- Implement anti-trust laws and other regulations that limit or prevent unfair market domination by corporations that sell seeds, animal embryos, agrochemicals, veterinary products and biotechnologies or/and process and distribute food and fibres
- Restructure the tax system to encourage employment and enhance agricultural biodiversity in the entire food and fibre production-distribution chain. Ensure greater public sector spending and fairness within the food system.

2

A Dynamic European Agricultural and Agri-foodstuffs Sector

An Increase in the GVA of the Agricultural Sector but a Reduction in its Share of GDP

Europe has seen GVA at market prices for agricultural products increase by a factor of 4.5 between 1970 and 1986 and by a factor of 1.2 between 1986 and 1996. By 1996 it had increased to ECU 145 billion, thus exceeding, since 1986, that of the United States. These results must be qualified for several reasons: GVA is expressed at current prices; the size of the EU has changed over this period with the accession of new countries in 1973, 1981, 1986 and 1995; and, lastly, European prices are higher than American prices.

The increase in the GVA of the agricultural sector in Europe is however more the result of volume than price. This has been especially true since the middle of the 1980s when prices tended to stagnate. Over the same period, productivity per hectare continued to increase. The rise in GVA in 1982 was attributable to a substantial increase in the volume of production, whilst the rise in 1989 related to an increase in the price of animal products on world markets. The United States saw similar trends in 1982 and 1989.

The decline in GVA recorded between 1991 and 1993 was linked to the reform of the Common Agricultural Policy, which tightened control on the factors of production (particularly by setting aside arable land) and substantially reduced intervention prices (cereals). This reform was accompanied by direct subsidies to producers in order to offset the fall in intervention prices. Thus, the proportion of direct subsidies in net income from agricultural activity was 37% in 1997 compared with 20% in 1992.

Since 1980, the average annual rate of increase in GVA has been 2%, for intermediate consumption it has been 3 %. The increase in GVA is closely linked to an increase in volumes produced, which in turn has led to the use of greater quantities of inputs.

In relative terms, agriculture, forestry and fishing have seen their share of GDP in the EU decrease over the last 26 years, falling from 6% in 1970, to 3% in 1986 and to 2% in 1996. This trend is closely linked with the rapid development of new branches of activity, particularly market services, which have grown in relative terms. Over the same reference period, the decrease was greater in Japan than in the EU. The United States presents a less clear picture, though the trend has tended to be downwards. For all three areas, the decline was particularly noticeable between 1973 and 1980.

Marked Disparities within the European Union

In absolute terms, the highest GVA for agricultural products in 1996 was achieved by France (ECU 28 billion) and Italy (ECU 27 billion), which together make up 38% of GVA in the EU for this sector. This is explained by the large volume of production (particularly in France) and the high value-added of agricultural products (particularly of Mediterranean-type produce in Italy). The greatest increase in GVA between 1970 and 1996 was recorded in the Netherlands, where agricultural activity increased more than fivefold. Over the last ten years, the greatest increase was in Greece, with GVA rising by a factor of 1.6.

On the other hand, expressing GVA in ECU per inhabitant gives figures for Greece, Denmark and Ireland which are more than double the Community average, whilst those for France and Italy are barely. Greece is quite different from the other Member States in that GVA for agricultural products accounts for 14% of GDP, i.e. more than six times the Community average. Income from agricultural activity is growing, but at a rate below that of the other sectors.

Income from agricultural activity, i.e. the compensation of the factors of production (capital and labour), can be estimated using agricultural GVA per agricultural job.

Between 1983 and 1996, agricultural GVA per agricultural job increased at an average annual rate of 4% (3% when jobs are expressed in AWUs), as a combined result of a sharp increase in GVA and a reduction in the number of agricultural jobs. There are wide variations in agricultural GVA per agricultural job within the EU. Denmark and the Netherlands, with their high value-added intensive agricultural activity, record very high levels of GVA per job.

A joint analysis of GVA and employment shows the intersectoral parity of incomes. Intersectoral parity exists when the gross value added (GVA) of the agricultural sector per agricultural job is equivalent to gross domestic product (GDP) per job. In the agricultural sector, GVA per agricultural job has been systematically lower than the average of all the sectors combined. One of the reasons for this is the size of the agricultural labour force and particularly the scale of part-time employment. Although agricultural productivity may not be very high, in certain regions of the EU agriculture is a key sector in maintaining employment levels.

Denmark, however, is an exception, with a unitary parity index. The lowest parity levels (index lower than 0.4) are those of Austria, Portugal, Germany and Luxembourg. This means that for these four countries, there is a considerable income disparity between the agricultural sector and the other sectors.

The Dynamism of the European Agri-foodstuffs Sector

In recent years the agri-foodstuffs sector in Europe has seen much more sustained activity than the agricultural sector. GVA at market prices for the European agri-foodstuffs sector increased sixfold between 1970 and 1986 and doubled between 1986 and 1996. The two significant increases seen in 1973 and 1986 were the result of two enlargements of the Union. However, this sharp increase in absolute terms corresponds to a fall-off in value as a percentage of GDP (3% in 1996 compared with 4% in 1970 and 1986). There are many reasons which explain this rapid development, namely:

- the establishment of major food groups, injecting economic dynamism into the sector;
- the changes which have taken place in marketing methods, with the development of large-scale distribution;
- the profound change in the lifestyle of consumers, who have become ever more demanding in terms of product quality and choice;
- the expansion and globalisation of international markets.

The same few countries leading the agricultural and agri-foodstuffs sector

In 1996 the highest GVAs in the agri-foodstuffs sector were in Germany (ECU 43 billion), the United Kingdom (ECU 34 billion), France (ECU 33 billion), Italy (ECU 23 billion) and Spain (ECU 19 billion). These five countries account for over 80% of the wealth produced in this sector in Europe. The same five countries are also the leading agricultural producers, although the order of ranking is different. Since 1970, Luxembourg, France and the Netherlands have made dynamic progress, recording above-average growth. The agri-foodstuffs sector in France is therefore not only of substantial importance to the national economy, it is also a leading growth sector. When GVA is considered on a per capita basis, Denmark leads the other Member States, followed by Ireland, Belgium and the Netherlands. A joint analysis of the shares of agriculture and agri-foodstuffs in total GVA clearly shows that these two sectors are very important for the economy of Portugal, Ireland and Greece. The agri-foodstuffs sector is more developed in Portugal, whereas Greece and, to a lesser extent, Ireland are more orientated towards agriculture. The United Kingdom, Belgium and Germany are more geared towards processing than towards production.

The European Union Remains a net Importer

The EU occupies a prime position in the world markets for agricultural and food products. Over recent years the volume of trade in unprocessed and processed products has seen a sharp increase. The EU internal market remains limited and these products have to find new openings in a strictly regulated world market. Intra-Community exports and imports represent 76% and 65% respectively of total trade.

The trade balance for these products is negative. This can be attributed to the high deficit in trade in unprocessed products (ECU 17.5 billion of exports compared with ECU 29.6 billion of imports). On the other hand, the balance in processed products has moved increasingly to the EU's favour. In 1997, exports reached ECU 1.2 billion, with imports at ECU 0.2 billion. Between 1989 and 1997, exports of processed agricultural products increased by a factor of 17 and imports by 6.

Analysing this trade by product shows that the main net exports of the EU are beverages-mainly wine-and, to a lesser extent, dairy products, cereal-based preparations and, lastly, meat and offal. On the other hand, the EU is a net importer of fruit and vegetables-amongst others, bananas and tropical fruit which are not produced in the EU-and of other tropical crops such as coffee, tea, cocoa and spices.

Varying Trade Balances Among Member States

As far as trade in 1997 in unprocessed and processed products is concerned, the EU can be divided into two groups:

- countries with a high trade surplus: the Netherlands [1], due to a large extent to exports of live plants and flowers; France with exports of beverages; and Denmark with exports of meat and offal;
- countries with a high trade deficit: Germany and the United Kingdom, with a very high demand for fruit and vegetables; and Italy, with imports of meat and offal.

The Netherlands (ECU 17.6 billion) and France (ECU 15.3 billion) are the largest exporters of unprocessed and processed products. Total imports exceed ECU 10 billion for four countries: Germany (ECU 18.1 billion), followed by the United Kingdom (ECU 12.4 billion), France (ECU 11.2 billion) and Italy (ECU 10.5 billion).

Agricultural Policy Incorporating Environmental Concerns and Rural Development

There is a broad consensus among agricultural and environmental economists that, as a general rule, the existance of a European agricultural policy which guarantees high prices over decades has contributed to the intensification and specialisation of agricultural production. The result has been negative effects on the environment, the landscape and the quality of produce, inter alia.

The reform of the CAP in 1992 marked a point of no return. The fixed prices for cereals and beef and veal were reduced and administrative guarantees of unlimited markets for agricultural produce were reduced or even eliminated, for example in the case of oil seeds, protein crops or beef. This trend continued with the reform of the organisation of the common market for fruit and vegetables and olive oil. This change in the CAP had two different effects on farmers' behaviour: one, an increase in 'reasoned agriculture' and the other, an intensification of production. Thus a more rational use of pesticides and fertilizers was seen. Use of inputs fell at first following the reduction in prices and the feeling of uncertainty in the agricultural world after the reform, but increased again when world and Community prices started to rise. It is still too early to see the effect of the change in price trends recorded in 1998 and 1999 on use of inputs.

The reform of the CAP in 1992 also encouraged the production of renewable raw materials for non food uses (bio-fuels, starch....), in particular on set aside land. In the same way, the use of biomass to produce energy represents a diversification for farmers and contributes to the creation of new jobs in rural areas. Among the measures which have potentially beneficial effects on the environment should be counted also the setting aside of land, whether voluntary or obligatory, if the system is properly managed.

Moreover, the reduction in cereal prices made intensive livestock rearing in cereal zones more attractive, and reduced the advantages of regions close to the ports. These regions had benefited from a certain logistic advantage, because of their privileged access to animal feedstuffs imported from third countries, thanks to an unbalanced protection tariff, to the detriment of cereals.

On the other hand, other elements of the 1992 reform had a negative effect on the environment. Sometimes this was the direct responsibility of measures adopted at Community level, and sometimes was due to the way in which Member States used the existing room for manoeuvre.

A good example of the first was the extension of cereal aid to silage maize; this was not proposed by the Commission but decided by the European Agriculture Ministers. This gave an added comparative advantage to intensive cattle rearing, particularly dairy cows, to the detriment of extensive livestock rearing.

Premiums for beef and veal are a second example of decisions by the Council differing from the proposals of the Commission. The Commission proposed that the special premiums for male bovines and suckler cows, excluding heifers, should be restricted by setting maximum stocking limits [2].

These are generally in extensive holdings and their meat does not have any problem finding a market. The final decision was to extend these premiums to the first two livestock units (LU) per hectare of the holding while continuing to exclude heifers, i.e. penalising extensive stock breeders. To compensate somewhat for this, an additional premium for extensification was introduced, for farms with less than

1.4 LU per hectare. The 1992 reform of the CAP introduced, at European level, support measures in favour of agri-environmental actions to promote more environmentally-friendly production methods. In this way incentives relating to:

- agricultural production methods compatible with environmental protection requirements-Council Regulation No 2078/92,
- the Community aid scheme for forestry measures in agriculture (Council Regulation No 2080/92),

have, despite limited budgetary resources, enabled organic farming to be promoted, involving 1% of all agricultural holdings in the EU (EU 15) in 1996, and more than half a million hectares to be reafforested in four years. The reduction in guaranteed prices did not produce a dramatic drop in farmers incomes, as predicted by many opponents of the reform. Net individual agricultural income increased by 4.1% per annum between 1992 and 1996 [3] (Source: Agricultural accounts), for the following reasons:

- direct compensatory aid provided for this effect;
- a high capacity of adaptation to change shown by farmers;
- world prices were maintained. Levies on wheat and durum wheat exports even had to be imposed several times during the 1995/96 and 1996/97 marketing period to ensure a sufficient supply for the internal market;
- a sharper than anticipated increase in cereal consumption in animal feed;
- currency devaluations which affected a number of Member States.

One can therefore conclude that from the point of view of the environment, the 1992 CAP reform represented a step in the right direction, even if several provisions had the opposite effect. In any event, it is now necessary to make more active progress towards the integration of environmental aspects and of sustainable development in the CAP.

It should not be forgotten that in terms of land occupation and landscape, European farmers manage and maintain 44% of the European area as Utilised Agricultural Area (UAA), and if the other areas that they own or rent are taken into account, they manage more than half of the European territory. Agriculture must therefore in no way be restricted to the sole function of economic production and the 2% of EU's GDP that it represents. Moreover, it has become more widespread for farmers to exercise other activities, with 15% of them (EU 15) employed part-time on their holding in 1997. New activities are emerging, thus creating sources of additional income. Agri-tourism and the processing of agricultural products on the farm have been developing for a number of years.

The proposals contained in the European Commission's Agenda 2000 confirm this orientation. Price reductions, reduction in intervention guarantees, increase in direct aid, modulation of aforesaid aid, strengthening of environmental obligations on agricultural producers, on one hand, and of positive agri-environmental measures on the other, rural development, make up the various instruments which, together, are intended to build a more coherent and effective integration strategy.

Changes in Agricultural Employment

Pierre Antoine Barthelemy (Eurostat)

There is a higher level of employment in agricultural production in southern than in northern Europe. Employment in agriculture is falling overall, although more slowly since the reform of the Common Agricultural Policy (CAP) of 1992, while the jobs situation in the agri-food sector is stable. Nevertheless, employment in both sectors is diminishing as a proportion of total employment. Farming is tending to become less important as a source of employment in rural areas even though it still plays a vital role in preserving rural landscapes. On the whole, agricultural employment accounts for a higher proportion of total employment than the proportion of GDP accounted for by agricultural GVA. This can be interpreted as meaning that structural adjustment is required, but it also reflects the importance of family labour and part-time work in farming. The falling number of farms and their increase in size does not necessarily go hand in hand with an increase in the number of paid workers.

A fall in agricultural employment and north-south differences.

In 1997, there were approximately 7.4 million people working in the agricultural sector in the EU (agriculture, hunting, forestry and fishing), of which 7.2 million were employed in agriculture proper (excluding fishing). On average, agriculture accounted for 5% of EU jobs in 1997, rising to over 10% in Greece, Portugal and Ireland, the three countries with full Objective 1 coverage.

The breakdown of agricultural employment across the EU must also be considered in terms of production types. In countries where the more labour-intensive Mediterranean-type production predominates (Italy, Spain, Greece and Portugal), farming accounts for 9% of jobs on average. Arable farming and animal husbandry, which are more common in the countries of northern Europe, require less labour (agriculture accounts for an average of 3% of employment in DK, NL, FIN, S and the UK).

The number of jobs in the agricultural sector is falling steadily. This is a major and continuing trend, linked to global economic developments and observable in all technologically advanced countries. The increase in the figures for 1986 was due to the accession of Spain and Portugal and the increase in 1995 to the accession of Austria, Finland and Sweden.

The fall in agricultural employment is affecting all Member States but in particular the countries with the highest levels of jobs in farming (Italy, Spain, Portugal and France). Each of these four countries lost more than one-third of its farming jobs between 1987 and 1997 (as opposed to one-quarter for the Community as a whole). Denmark also appears to have suffered heavy job losses but this is not actually the case. In reality these apparent losses have been due to the development of parallel structures, with employment in these structures being classified outside the agricultural sector. Belgium, on the other hand, shed only 5% of its agricultural jobs over the ten years in question.

From the year 1992 the downward trend has slowed: Following the CAP reform the rate of disappearance of farms has noticeably reduced. From – 5.2% in 1991 it has steadily shifted to be –1.6% in 1998 at the Community level.

There are differences between northern and southern Europe in terms of the age distribution of heads of holdings. The proportion of elderly farmers is generally higher in the Mediterranean countries, with nearly one in two farmers over 55 years of age as opposed to only 1% in Germany. Just 4% of Portuguese farmers and 6% of Italian farmers are under 35 (the Community average is 10% under 35).

In relative terms, employment in the agri-food sector is highest in Denmark and Ireland : In 1997, there were 3.5 million workers in the food and tobacco industries in the fifteen EU countries. In absolute terms, Germany, France and the United Kingdom topped the list with almost 60% of the EU's agri-food jobs. While Italy and Spain had the highest number of jobs in agriculture, they came in behind the others in terms of agri-food jobs. These five countries together account for almost four out of every five jobs in the European agri-food sector.

The agri-food sector accounted for approximately 8% of EU industrial employment and 2% of total employment in 1997. In relative terms, Ireland and Denmark were the biggest employers, with the agri-food sector accounting for more than 3% of all jobs. The proportion was much lower than 2% in Luxembourg, Sweden and Italy.

The trends in employment in the agri-food sector vary greatly. Between 1987 and 1997, employment levels increased in some countries. This was the case in France, and also in Greece, the Netherlands and Portugal (even though the numbers remained low there). On average, the number of jobs remained static or even diminished slightly. The drop was more marked in Luxembourg, Denmark and the United Kingdom. In general there is a strong correlation between the fall in the number of agri-food jobs as a proportion of total employment and the drop in employment in the industrial sector overall, which is linked to productivity gains achieved over the decade.

During the 1980s, company mergers and takeovers radically transformed the structure of the agri-food sector, with multinationals accounting for a growing share of the market. However, in numerous market segments, the degree of concentration is still limited, given the wide diversity of European agri-food products and the existence of regional patterns of consumption. A large number of SMEs therefore contribute to employment in the sector.

Structural Weakness in the Agricultural Sector

A joint analysis of the levels of agricultural and agri-food employment highlights the special situation in Ireland, Greece and Portugal, where agriculture plays a dominant role in terms of processing and marketing jobs. This is also true to a lesser extent in Spain, Austria, Finland and Italy.

Changes in the number of agricultural jobs do not correlate with changes in the number of agri-food jobs. The countries where agricultural employment has fallen sharply are different from the ones where agri-food employment has fallen. Denmark is the exception, however; for reasons already mentioned, it has experienced a fall in the number of jobs in both sectors.

Employment in agriculture is generally higher than agricultural GVA in the Member States in percentage terms, with the exception of Denmark (where it is lower), and the Netherlands and the United Kingdom (where it is almost the same). The gap between the two variables is particularly wide in the case of Portugal (where agriculture accounts for 13% of jobs and 4% of GDP) and Ireland (11% of jobs and 5% of GDP).

This comparison highlights both the possible need for structural adjustment in the agricultural sector in certain regions and the importance of farming jobs in some rural areas. This view is supported by other variables related to the structure of holdings, such as the average SGM per AWU and the proportion of the total agricultural labour force working on small holdings (holdings of less than 8 ESUs). Thus, Portugal has the lowest average SGM per AWU (4 ESUs compared to the Community average of 20 ESUs) and the highest proportion of its agricultural labour force working on small farms (74% compared to the Community average of 38%).

Two agricultural jobs and one agri-food job per hundred inhabitants

There were on average twice as many jobs in agriculture as in the agri-food sector in the EU in 1997. The ratio has been falling since 1983, and there are major variations across the Member States. At opposite ends of the spectrum are the United Kingdom (with a ratio of 1:1) and Greece (with a ratio of 7:1).

There were on average two agricultural jobs and one agri-food job per hundred EU inhabitants in 1997. The number of farmers per hundred inhabitants has tended to fall while the number of agri-food jobs per hundred inhabitants has remained static. Portugal lost four jobs per hundred inhabitants in the two sectors combined between 1987 and 1997, while Greece and Spain lost two.

Family Labour is the Keystone of Agriculture

Family labour predominates in agriculture, accounting for four out of every five jobs in 1995. The number of non-family workers is highest in the United Kingdom and Denmark. Elsewhere, farming is still a family affair. In Finland, for example, family members account for over 97% of the agricultural work force.

The fall in the number of farms and their increase in size has not necessarily led to a corresponding increase in the number of paid workers. These changes have affected the Member States in different ways. In countries such as Denmark and Greece, there has been a large increase in the number of paid workers and the same is true to a lesser extent in Belgium, Spain, the Netherlands and Luxembourg. These changes are linked to gains in competitiveness and productivity, which are achieved by:

- increasing production capacity;
- rationalising existing farm structures;
- developing labour-intensive agriculture with higher value-added, such as fruit and vegetable growing and, more generally, Mediterranean-type production;
- developing alternative activities.

There has been a fall in both the number of farms and the number of paid workers in Germany, Portugal, Italy, and, to a lesser extent, Ireland and France.

Throughout the EU, the reduction in the number of farms has entailed lay-offs of paid workers, as in every other sector of the economy. This is not the only impact of this reduction, however. Family workers too are swelling the ranks of the unemployed.

Women make up one-third of the total agricultural labour force in the EU. The proportion of female farm workers is particularly high in Portugal and Austria (52% and 49% respectively).

Farmers are becoming better educated. More than one farmer in ten has a higher education qualification in Germany, the United Kingdom and Ireland (17%, 11% and 10% respectively), while the Community average is 6%.

The OECD uses the educational level of farmers as an indicator in its work on agri-environmental indicators, on the basis that it is generally agreed that the higher their level of education, the more concerned farmers are with environmental issues.

Part-time Work Predominates in Southern Europe

Part-time work is much more widespread in the countries of southern than northern Europe. In Greece, Portugal, Italy and Spain, nearly half of all farmers (compared to less than 30 % in the other eleven Member States) and more than half of all agricultural labourers work part-time. In Italy and Greece, more than one in three farmers spend less than half of their working time on the farm. The large number of seasonal activities in southern Europe is one explanation for this high level of part-time work.

Part-time work may mean that full-time work is impossible or that the farmer is involved in other non-agricultural gainful activity. Such activity is common in Ireland and Germany, with more than one farmer in four engaged in a non-agricultural gainful activity, but not so in Luxembourg or Belgium. The non-agricultural activity is the main job everywhere except in France, which implies that the farming is done on the side, possibly as a hobby.

A joint analysis of the proportion of part-time jobs and the proportion of farmers engaged in a non-agricultural gainful activity highlights the special situation in the southern European countries. While there is a high level of part-time employment, very few farmers are engaged in a non-agricultural gainful activity. There would appear to be underemployment, perhaps due to structural weaknesses in the agricultural sectors in these countries.

Agriculture Remains a Driving Force for Economic and Social Cohesion

Overall, agricultural employment is clearly falling in Europe. This does not mean, however, that agriculture is doomed to play a secondary role in the process of economic and social cohesion in certain regions, and particularly rural regions. Even as a minority in the countryside, farmers are still the main managers of the land and agricultural work largely determines the degree of attractiveness of these regions, particularly where the landscape is concerned.

The EU has long been aware of this basic role of farming and has encouraged certain diversification measures. Two Regulations in particular have been instrumental. Council Regulation (EEC) No 2078/92 of 30 June 1992 concerns agricultural production methods compatible with the requirements of the protection of the environment and the maintenance of the countryside. Some of its objectives are covered by an aid scheme to promote:

- the upkeep of abandoned farmland and woodlands where this is necessary for environmental reasons or because of natural hazards and fire risks, and thereby to avert the dangers associated with the depopulation of agricultural areas;
- long-term set-aside of agricultural land for reasons connected with the environment;
- land management for public access and leisure activities.

Similary, Council Regulation (EEC) No 2080/92 of 30 June 1992 instituting a Community aid scheme for forestry measures in agriculture is intended to promote afforestation as an alternative use of agricultural land and the development of forestry activities on farms.

A significant number of diversification activities have been developed in recent years. In a given area these activities are a new factor in the social integration of economically-active persons. Four fields appear to be of particular importance in terms of job creation:

- preservation of the landscape, protection of areas of great natural value and habitats such as wetlands, rivers and protected areas;
- sustainable management of forests to preserve biodiversity and fulfil other functions (e.g. of a recreational nature);
- the development and use of plants for non-food purposes, e.g. in industry or energy production;
- organic farming.

Other areas with job-creating potential, particularly in rural regions, are green tourism, the management of heritage resources, the supply of services to the community, etc. To exploit these opportunities, farmers need to mobilise regional and local authorities, businesses, NGOs and funding bodies. Their involvement in environmental management can promote job-creation and generate supplementary or alternative sources of income. In order to meet these new challenges, the farming sector has already begun to adapt in some regions, in particular through the emergence of new forms of employment (partnerships, employers' groups, etc.). It is impossible at the present time to quantify these activities precisely.

The political challenge that Agenda 2000 tries to meet in its part on agriculture and particularly rural development, is to weaken or even break the link between leaving agriculture and leaving country. The new rules give much more freedom to national or responsible authorities to mount programmes tailored to the local situation. At the express request of the Council of Ministers the application of the new rules has been enlarged to cover "diversification of activities in or close to agriculture" and "basic services required by the rural economy and rural population".

Environmentally Friendly Farming Methods Generate Employment

It is difficult to quantify the impact of more environmentally-friendly agricultural production on employment, but a fairly clear picture is provided by organic farming, the most highly developed and structured branch of sustainable farming. While no exact figures are available, it can be said that, in order to meet consumer demand, this branch of agriculture must be clearly organised to:

- ensure the credibility of this type of production, and
- open up consumption channels in order to move from being a niche market to capturing a substantial market share.

In order to meet these requirements, a number of structures are needed. To monitor compliance with the rules on organic production, certification bodies – which provide jobs-have been set up in the Member States. Organic farming also requires technical and economic know-how for which specialist technicians are needed. Their technical advice is a key factor in helping farmers to decide whether to move from traditional farming to organic farming. With regard to consumption channels, organising the collection of produce also generates new jobs.

In France, for example, specialised cooperatives organise some of the collection of organic cereals at regional level. Furthermore, the move by major food retailers into the market for organic produce has forced organic farmers to restructure in order to ensure regular supplies of produce in sufficient quantities. In Austria, a group of organic producers called *Ernte* enters into contracts directly with major food retailers via a firm called *Ökoland*. Although the downside of this is that producers lose their independence to some extent, this mode of organisation boosts organic farming and creates employment, because supermarkets increase the market share by selling organic produce to a new category of consumers.

In Great Britain agri-environmental measures are leading to a small increase in agricultural work and a substantial increase in work by firms. The activities of hedging, maintenance of terraces and repairing stone wall involve intensive work needing extra labour. Similarly in the Netherlands livestock farmers who have signed management contracts need additional labour. In Denmark converting farmers to organic farming leads to increasing labour cost: these rise significantly (38%) when setting up organic farms. This result can be taken as a positive employment effect. Finally, although job creation is not the objective of agri-environmental programme, nonetheless results confirm that improved environmental management of the land requires additional labour. Developing the concept of environmental services provided by farmers affords new job opportunities.

Soil at the Interface between Agriculture and Environment

Agriculture and Soil Conservation

Unlike air, water, and biota, which are mobile systems, soil is site-specific, and although more stable than the other three systems, it shows great variability in space and time.

Soils have at least six main functions relevant to human life:

- The production of biomass by agriculture and forestry.
- Filtering, buffering and transformation activity, between the atmosphere, ground water, and the plant cover, protecting the environment and especially humans through the protection of the food chain and drinking water reserves.
- Soils are biological habitats and gene reserves, much larger in quantity and in quality than all the above ground biomass.
- Soils serve as a spatial base for technical, industrial and socio-economic structures and their development, e.g. for the construction of industrial premises, houses, transport systems, sports and recreation areas, dumping of refuse, and others.
- Soils are used as a source of raw materials, e.g. clay, sand and gravel for construction, and also as a reserve of water and energy.
- Finally, soils are a geogenic and cultural heritage, forming an essential part of the landscape in which we live and containing palaeontological and archaeological treasures of high value for the understanding of the history of earth and mankind.

The problems of soil degradation and soil destruction are caused by the competition between these different forms of land use. Therefore, new perceptions and concepts for sustainable land use should be developed, which are in conformity with the constaints of nature. In this context, sustainable land use and protection of soil can be defined as the spatial (local or regional) and temporal harmonisation of all main uses of soil and land, minimising irreversible effects. This is a political rather than a scientific issue.

Soil is affected by physical, chemical and biological degradation. Some agricultural activities contribute to these negative effects. However, it must borne in mind that industry, urbanization, road construction, fire, other human activities and, more generally, demographic pressure and climate changes are also major factors.

On the other hand, good agricultural practices, geared towards the conservation of the major soil functions, are well documented. Traditional farming practices ensured in many parts of Europe, especially in the Mediterranean region, the long-term conservation of soil fertility. Erosion control measures, integration of organic matter, crop rotation, etc. were all traditional farming practices well known to the European farmers.

The European Union, through its Common Agricultural Policy, has well recognized the importance of such beneficial agricultural practices. An agri-environmental incentive plan, based on regulation 2078/92, plays an important role in preserving these agricultural practices, which tend otherwise to be abandoned in favour of more remunerative, but less sustainable, technologies.

The most significant forms of *physical degradation* of the soil due to agriculture are:

- erosion;
- desertification;
- water-logging;
- compaction.

Land use practices such as deforestation, overgrazing, some agricultural cultivation practices, removal of vegetative cover or hedgerows can exacerbate these occurrences. The increasing demand for water, the sometimes excessive mechanization and ploughing are further causes of such degradation.

The following processes characterize *chemical degradation*:

- acidification;
- salinization;
- contamination by micro-pollutants, such as pesticides and their metabolites, heavy metals and nutrients *i.e.* nitrogen and phosphorous. (However, some pesticides may stay in the soil for some time without any consequences for the environment).

The major consequences of the above contamination are toxification and eutrophication.

Related agricultural practices are:

- over-use of manure and mineral fertilizers;
- emissions of pollutants by intensive livestock production;
- spreading of sewage sludge on agricultural soils;

- the use of pesticides with unintended side-effects (slow degradation).

As regards *biological degradation*, it should be remembered that the quality of the soil is mainly defined by its biological activity, which is affected by:

- important humus mineralisation; and
- changes in biodiversity.

Lowering the humus content makes soil more susceptible to compaction, erosion and other forms of physical degradation. Inappropriate land use practices, especially in agricultural fields, are most often the reason for this problem.

The unintended side effects of pesticide use on soil life can explain many changes in biodiversity. However, this occurrence must be considered in conjonction with the degradations described above.

Environmental Issues and Trends Regarding Soil use in the EU

Sealing and Urbanisation

The current rate of soil loss by sealing through urban expansion and infrastructure in the Netherlands amounts to ca. 36 ha per day. In other countries of Europe, like Germany (120 ha per day), Austria (35 ha per day) and Switzerland (10 ha per day), similar soil losses by sealing through infrastructure and buildings have been observed. This urban sprawl increases the costs of urban infrastructure, traffic in urban areas, and energy consumption, and has negative effects on the quality of the countryside and the environment. This development is in direct competition with agricultural land uses and is threatening valuable agricultural soils all over Europe.

Soil Erosion

According to the Dobris+3 report, about 115 million hectares of European soils are suffering from water erosion and 42 million hectares from wind erosion.

Soil erosion is a major socio-economic and environmental problem through Europe. It reduces the productivity of the land and degrades the performance and the effectiveness of the ecosystems. More than half of the land in Europe has suffered various degrees of soil erosion by water and about a fifth has been eroded by wind. The phenomenon is more acute in the southern countries, where it often reaches catastrophic dimensions, than in the northern territories.

Soil erosion and salinisation have increased the risk of desertification in the most vulnerable areas, particularly in the Mediterranean region. Information on the extent and severity of desertification is limited; further work is needed on prevention strategies, possibly within the framework of the United Nations Convention to Combat Desertification (UNCCD).

Water Erosion

The importance of water erosion in Europe, in terms of area. The dominant effect is the loss of topsoil. That is often not conspicuous, but nevertheless potentially very damaging. Physical factors like climate, topography and soil characteristics are important in the process of soil erosion. In part, this explains the difference between the severe water erosion problem in Iceland but the virtual absence of erosion in Scandinavia where the climate is less harsh and the soils are less erodible. Comparison can be made between the erosion occurring on the coasts of the northern Mediterranean, from Spain to the Adriatic, and erosion occurring in North West Europe (Netherlands, N. Germany, Denmark). The Mediterranean region is subject to long dry periods followed by heavy bursts of erosive rainfall, falling on steep slopes with fragile soils, resulting in considerable amounts of erosion.

This contrasts with NW Europe where soil erosion is slight because rain falls on mainly gentle slopes and is more evenly distributed throughout the year. Consequently, the area affected by erosion is much more restricted in its extent. In parts of the Mediterranean region, erosion is now irreversible and in some places has practically ceased because there is no more soil left!

The natural formation rate for new soil is between 0.1 to 10 tonne/hectare/year. With such a very slow rate of soil formation, any regular soil loss of more than 1 t/ha/yr can be considered as irreversible within a time span of 50-100 years. Losses of 20 to 40 t/ha in individual storms, that may happen once every two or three years, are measured regularly in Europe with losses of more than 100 t/ha in extreme events. It may take some time before the effects of such erosion become noticeable, especially in areas with the deepest and most fertile soils or on heavily fertilised soils.

However, this is all the more dangerous because, once the effects have become obvious, it is usually be too late to do anything about it. The major causes of soil erosion are still inappropriate agricultural practices, deforestation, overgrazing and construction activities. The problem is greatest in the Mediterranean region because of its fragile environmental conditions, but problems exist in most European countries. Land abandonment and forest fires, particularly in marginal areas, intensify soil erosion. Strategies, such as afforestation, for combating accelerated soil erosion are lacking in many areas.

Wind Erosion

The dominant result of wind erosion in Europe is loss of topsoil. It occurs mostly in south-eastern Europe. The removal of soil material is selective, even more so than for water erosion, as only fine particles (clay, organic matter with their adsorbed nutrients) are removed, leaving impoverished, coarse textured soil particles.

The off-site effects are increased air pollution because of the presence of dust particles, the covering of fertile soil layers and, in the most extreme cases, complete burial and dune formation. Terrain deformation in the form of windblown depressions may also occur as well. Damage through wind erosion is practically irreversible.

Compared with water erosion, wind erosion shows a more distinct regional distribution in Europe. This suggests that physical factors, in particular climate, are relatively more critical than human influences.

The fairly moist conditions prevailing in Western Europe and the nature of storm events, often associated with rain from oceanic depressions, diminish the chances of wind erosion.

This is not a universal rule, however, demonstrated by the incidence of severe wind erosion in Iceland and the less severe, but still significant, events in certain parts of North West Europe (N. Germany, the Netherlands, E. England).

Extensive conservation measures such as windbreaks have considerably reduced the hazard. The widespread and serious wind erosion in the south eastern Europe (including the S. Russian plain) can be partly explained by the dry continental climate and sensitive soils in combination with inappropriate farming practices.

Excessive drainage is a major cause of conditions that favour wind erosion. However, human activities again play their part in causing wind erosion. Overgrazing appears to be a major cause, in part because the areas most sensitive to wind erosion (semi-arid regions with sandy soils) are less suitable for other types of land use.

The Nordic Countries

The information available on soil erosion in the Nordic countries (i.e. Norway, Sweden, Finland and Denmark) was recently compiled in a report, edited by Rekolainen and Leek (1996).

The report tries to assess the erosion risk over the whole territory of these countries. In Norway, Sweden, Finland and Denmark, water erosion is considered to be the main problem, not only because of the negative impact it may have on soils and agriculture but also because of its significant contribution to the phosphorus loading of freshwater bodies.

The water erosion risk was assessed using a model based technology (a core model, USLE and a supporting model CREAMS for some, more detailed analyses). The necessary input data were derived from existing data sets on climate, soils, crop type and topography and the results presented in the form of maps.

When compiling the maps, it appeared necessary to adjust the modelling strategy to the available data in each country. This is probably the reason why the final results are presented in a relative form, with 5 classes of the relative erosion risk, although all calculations are based on a quantitative model.

The maps covers the whole territory of Finland, Sweden and Denmark. In Norway, the necessary topographic data were only available for *ca.* 25% of the total territory.

The following areas have a high risk of erosion: southern Finland; eastern Jutland, Seeland and Fünen in Denmark; the area around Oslo in Norway; the southern coast and the area between Göteborg and Arvika in the west and Västervik and Gävle in the east in Sweden. The qualitative nature of the documents makes it impossible to assess the severity of the problem in absolute terms.

The Central-Western Countries

Soil erosion by water in Central Europe is not as aggressive as in Southern Europe, because bioclimatic, topographic and human pressures have not been as acute as in the south (Yassoglou, 1987).

The bioclimatic conditions in Western Europe also help avoid high rates of soil erosion. Favourable evapotrans piration rates and the even distribution of rainfall throughout the year facilitate the speedy regeneration of the protective vegetative cover of the land.

Low relief and gentle slopes over large parts of the region contribute to the low rates of erosion, which average 0.24 t/ha/yr. In the mountainous regions of France, the losses range from 1.8 to 2.5 t/ha/yr whereas in small valleys in the Alps and the Apennines losses reach 25 t/ha/yr.

However, the potential erosion risk, arising from removing the protective vegetative cover by cultivation, is significant even on the gently sloping land and considerably exceeds the tolerance level.

In Belgium the area susceptible to water erosion is estimated at about 10 per cent of the agricultural land area, while improvements in artificial drainage have reduced the problem of waterlogging.

Rates of soil loss in Belgium have been reported as high as 82 t/ha/yr., from bare fallow land on 5-7% slopes. The dominant form of soil erosion in France is water erosion, estimated to affect about 5 million hectares of agricultural land, or about 17 per cent of the total, compared with 0.5 million hectares at risk from wind erosion, mainly in the south of the country.

Soil erosion now effects most of the main cereal growing areas in France and other major agricultural production regions. This compares with the 1950s when it is estimated that 2.7 million hectares were affected by erosion, or 8 per cent of total agricultural area. About 37% of the arable land in England and Wales show erosion rates above the tolerance value. In Germany, potentially erodible surfaces are those on hilly sites with slopes as low as 2-6%.

Many data exist in the countries of Western-Central European, but the regional assessments of erosion risk are qualitative and based on a variety of methodologies.

As a pilot exercise, a semi-quantitative assessment for France has been made by INRA in collaboration with the European Soil Bureau (Kirkby *et al.* 1998). This approach could be applied to the whole continent and with local validation could provide a Pan-European soil erosion risk assessment map.

In Spain, soil erosion in certain olive orchards on steep slopes has been measured at up to 50 tonnes per hectare. In total, erosion and desertification in this country are estimated to severely affect (soil losses of over 100 tonnes/ha/year) as much as 9 million hectares or 18 per cent of the total land area, with 13 million hectares (26 per cent) moderately affected (12-50 tonnes/ha/year).

The factors that most influence vulnerability to erosion are steep slopes, drought followed by intense rainfall, lack of topsoil and sparse vegetation cover. A major part of the land affected by erosion in Spain is non-irrigated grassland, the so-called maquis or garrigue, and to a lesser extent areas of permanent crops such as vines, almond and olive groves.

The cost of the direct impact of erosion-including the loss of agricultural production, impairment of reservoirs and damage due to flooding-is estimated by the EEA at 280 million ECU per year, with the cost of soil rehabilitation estimated at about 3,000 million ECU over a period of 15 to 20 years.

Mediterranean Europe: The Corine Assessment

In the Mediterranean, the climate, the strong relief (characterised by steep slopes and extensive dissection), and the long history of human intervention (in the natural ecosystems) have resulted in rates of soil erosion, that are more catastrophic than in other regions of the Continent.

According to CORINE (1992), about 66% of the rural area has a moderate to high potential risk of soil erosion (by water). The distribution of the risk in the region, and within the individual countries, is complex. A large proportion of land in Portugal, Greece and Spain (68%, 43% and 41% respectively) is at high risk from soil erosion, whereas the corresponding proportions in Italy and in Mediterranean France are less (27% and 9% respectively).

High-risk areas tend to occur in the mountains of Cantabria, the Sierra Nevada, the Pyrenees, the Alps, the Apennines and the Pindos range. This reflects the control exerted by topography on slope and climate. High risk areas associated rugged terrain with subject to high climatic erosivity and soil erodibility also occur throughout Portugal, in Extremadura and Andalucia in Spain, in Corsica, in Sardinia, Calabria and Sicily in Italy, and in Crete and the Aegean Islands in Greece.

Extensive areas of bare rock cover approximately 10% of the land in Greece, Albania, and the Mediterranean sections of the former Yugoslavia. These areas can be considered to be mostly desert now and soil erosion has ceased to take place because there is practically no soil left. However, these areas still produce rapid surface runoff when it rains and this threatens lower lying land.

The area of actual erosion risk given the present vegetative cover, is considerably smaller than the area of potential risk and is estimated to cover 30% of Portugal, 29% of Spain 1% of Mediterranean France, 10% of Italy and 19% of Greece.

Salinisation

Soil salinisation affects nearly 4 million hectares, mainly in Mediterranean and Eastern European countries. The main causes are over-exploitation of water resources as a result of irrigation for agriculture, population increase, industrial and urban development and the expansion of tourism in coastal areas.

The main effects in cultivated areas are lower crop yields and even total crop failure. Strategies to combat soil salinisation are lacking in many countries.

3

Effects of Biotechnology on Agro-biodiversity

Since the *Convention on Biological Diversity* (CBD) was signed in 1992, the value of biological diversity in general, and agro-biodiversity in particular, has gained recognition worldwide. Agro-biodiversity can be defined as the total of components, structure and functions in agro-ecosystems relevant for agricultural production.

Agro-biodiversity is first of all of vital importance for the food security of future generations. This diversity can be exploited to overcome new pests and diseases, to cope with climate changes and a growing world population, to react to changing consumer demands and to make production more sustainable. Furthermore, agro-biodiversity is related to cultural values and traditions. Finally, many believe that diversity has an intrinsic value and that mankind is responsible for the survival of ecosystems and all that constitutes these systems.

Technology in general has had an impact for ages on the development of agro-biodiversity. Utensils in agriculture, food processing technology and seed storage systems, for example, have always been developed in close interaction with the diversity present in agriculture.

This includes the genetic diversity of landraces and varieties within crop and animal species, the diversity of different crops and animals at the farm and the diversity within the entire agro-ecosystem. Some technologies, like fermentation technology, have influenced the properties that farmers consciously selected for their crops and animals.

Others, like seed storage and fodder production, have resulted in conscious as well as unintended selection in landraces and animal breeds. The adaptation of crops and animals to the available technologies has determined their role in the farming system. And finally, the development of the farming systems has greatly influenced the agro-ecosystems in which they function, as is evident from the current discussions on the sustainability of high-external-input systems. Biotechnology, like all preceding technologies, has influenced and will further influence agro-biodiversity. In this article an inventory will be made of the likely or potential effects of modern biotechnology on agro-biodiversity.

Common Effects of Agriculture and Biotechnology on Biodiversity

Both agriculture and biotechnology make use of biodiversity and thus show us the value of biodiversity. Likewise, both agricultural practices and biotechnology impose a threat to the existing biodiversity on which they depend. Modern agricultural practices, stemming from the rise of a modern breeding industry and from the Green Revolution, have caused massive genetic erosion, the disappearance of many diverse populations of crops maintained by farmers and adapted to local circumstances.

The application of modern biotechnology may result in a wider use of genetic diversity, whether present in wild or domesticated species, for the benefit of future food security. However, it may also potentially result in the further narrowing of the genetic base of our food crops, because of the high costs of biotechnology and, consequently the tendency to focus on few varieties or breeds only. It may even result in the introduction of novel organisms which form a risk to the (agricultural) environment.

Genetic erosion, now commonly regarded as a negative development, was caused by the rise of the modern breeding industry and marketing decisions of that industry, and did not form a natural necessity. The socio-economic context in which biotechnology is developed and applied includes competition, research structures, seed markets and intellectual property rights. It will also determine which applications and, consequently, which effects of biotechnology on the conservation and utilization of biodiversity will dominate. For a number of biotechnologies the picture of positive and negative potentials, depending on the socio-economic context, begins to appear.

In Vitro Technology and Reproduction Technology

Tissue culture of crops offers the possibility of rapid propagation of large numbers of genetically identical plants. The technology can be used for the propagation of desirable genotypes of vegetatively propagated plants.

It is applicable in particular in the case of highly heterozygous clones, of which the genotypes are lost in sexual reproduction, and for plants with decreased fertility or long generation cycles. Tissue culture is therefore popular in horticulture and forestry. The technology can also be used to lower the virus pressure in starting materials. These applications may increase the availability of genetically diverse starting materials. Moreover, tissue culture can be used for conservation of genetic diversity. However, if the technology is mainly used to supply large areas with genotypically identical materials, the net effect of the technology on genetic diversity will be negative. At present, the net effects are still difficult to measure. However, for some crops which have already been reproduced by conventional clonal propagation techniques for many years, the trend is clearly negative. In the case of banana, only a few varieties are dominant worldwide.

Inbreeding of properties from distant relatives using in vitro methods can result in novel combinations of genetic diversity. In practice this option is not very popular in breeding because of the long generation times and high costs related to the introduction of genes from distant relatives. It has mainly led to rarities like a hybrid between potato and tomato and the 'shoat', a hybrid of sheep and goat.

Cryopreservation, the preservation at temperatures as low as -196 °C, can be used for conservation of plant meristems, but until now by far the most important application has been the conservation of sperm of animal species. Effects on genetic diversity of this application in artificial insemination have been mainly negative: the very successful bull Sunny Boy has produced the sperm for nearly 1 million inseminations in the Netherlands, which is over 75 per cent of the total, and left a massive genetic impact on Dutch livestock.

Multiple ovulation, in vitro fertilization and embryo transfer in combination with cryopreservation may allow the rapid propagation of selected animal breeding material. This results in the rapid replacement of heterogeneous, locally adapted populations by homogeneous stocks, as has happened in the case of chicken.

Apomixis and Terminator Technology

Apomixis, the seed formation from unreduced egg cells without fertilization, occurs in nature in some plant species like cereals. The resulting offspring is genetically identical to the parent. For commercial breeding this development is of little interest since it may create a way to surpass the biological property protection offered by hybrid varieties. It offers farmers the possibility to produce seeds for the next growing season with the same properties as the parent plant, which until now cannot be realized using hybrids.

However, modern plant-breeding industry may move into crops largely ignored so far, using a new technology, called terminator technology. The American cotton seed company *Delta and Pine Land* and the *United States Department of Agriculture* (USDA) announced they had received a patent on a technique that genetically disables a plant to set seeds that germinate when planted again. This technology can potentially be used in all cultivated crops.

Varieties of crops like rice, wheat, sorghum and soybean, which could not effectively be marketed using hybrids may now be commercially protected by using this new technology. This would open some of the world's largest food crops to profitable breeding and further increase the risk of genetic erosion, which has already progressed extensively in these crops. The net effect of this technology is likely to be negative.

Biological Test Kits and DNA Marker Technology

Using biotechnology, ready-to-use and simply applicable test kits have been developed for identification of viruses, fungi and other pathogens in the field. These kits can be implemented as precision instruments for a rational selection of control strategies and moments in integrated pest management. As compared with traditional pesticide use, integrated pest management has a positive effect on the total biodiversity in agro-ecosystems. Similar kits applied in diagnosis of animal diseases may remove barriers between nations in the exchange of breeding material for livestock and this could also increase the availability of a wider array of breeding material. The likely net effect of biological test kits on agro-biodiversity is therefore positive. Many agronomically important traits are governed by a large number of genes which interact at different regulatory levels. These traits are called quantitative traits. Traditionally, manipulation of the genes involved forms the domain of breeding.

Molecular markers enable localization of the genome sites involved in such properties and this knowledge may render breeding faster and more precise. Marker technology can be applied for the exploitation of genebank ex situ collections as well as for the improved use of elite breeding materials. It can be used for quantitative traits such as yield, mineral use, food quality, taste, storage capacity, resistance and adaptation to physical stresses.

Considering that most commercial crop varieties currently exhibit a narrow genetic basis, marker technology offers the opportunity for rapid corrective measures in case of outbreaks of new pests and diseases. Furthermore, the technology can be used preventively to determine the relationship between genetic materials and genetic properties and enables rational choices aiming at an increase of genetic diversity. In animal breeding determination of genetic relationships in animal populations is a major application. This allows a rational selection of reproduction schemes which best maintain the genetic diversity present in the population.

Finally, the technology can be used for fundamental research on the relationship between the level of genetic diversity in a crop and its versatility in reaction to changing conditions. Marker technology will surely be used by industry to improve current breeding practices, but it is uncertain to what extent applications will be realized that benefit maintaining or increasing biodiversity and versatility in agriculture.

Genetic Modification

Genetic modification has been applied in both plant and animal breeding, although commercial applications have only been realized in the former domain, e.g. insect resistance and herbicide resistance in maize, cotton and soya bean, and prolonged shelf life of tomato. In view of the high costs of genetic modification, the technology will only be applied in a few economically important crop varieties of major crops grown in the industrialized world, and possibly in animal breeds.

If the resulting genetically modified organisms form an agricultural success this may lead to replacement of several conventional varieties and thus to a decrease in genetic diversity in that crop or animal. Furthermore, patent protection will limit the use of such varieties in further breeding by others and thus prevent the spread of the newly introduced traits. Although patent protection is not a necessary consequence of biotechnology, its result is that the net effect of genetic modification on biodiversity is likely to be negative. This is contrary to popular statements about the addition of novel genes to available crop genotypes.

Genetic modification of micro-organisms with an agricultural value such as genetically modified Bacillus thuringiensis may improve bio-insecticidal sprays or enlarge their host range. Modification in relation with nitrogen fixation, which can reduce the use of artificial nitrogen fertilizers, offers only very limited opportunities in the short run because of the intricate biochemical pathways involved. Positive effects on biodiversity stemming from the reduction of the use of artificial fertilizers is therefore unlikely.

Genetically modified veterinary vaccines offer an opportunity to increase the health of all sorts of livestock, be it indigenous or exogenous, common or rare and net effects are hard to predict. However, increased use of more effective vaccines might fit the tendency to control the environment rather than to exploit genetic diversity in the search for better production. A positive role of biotechnology in maintaining and enlarging agro-biodiversity. Is the picture of the effects of biotechnology on agro-biodiversity indeed so gloomy? Much will depend on the ways biotechnology is exploited, and positive applications are certainly feasible. To summarize, what is the potential of biotechnologies to stop and reverse the decrease in diversity in agriculture?

Maintenance of genetic diversity. Marker technology can be used to study relationships between genetic material, in starting materials for breeding and in small animal populations, and to monitor the genetic developments in situ and on-farm conservation projects. Cryopreservation increases the options for conserving starting material, in particular for animals and for those crops which cannot be maintained using normal practices.

Utilization of genetic diversity. Markers can also be used to increase the efficiency of breeding strategies and to focus on desirable traits. Since markers allow precise identification of the genetic make-up of varieties, they allow relaxation of legislation on uniformity of crop varieties and may support reintroduction of genetically more heterogeneous and versatile varieties. Finally, markers can be used to trace properties which are important in ecological farming or other forms of sustainable production which are currently under-utilized. Cryopreservation in combination with artificial insemination may help to support the maintenance of rare breeds or rare properties.

Reduction of pesticide use. Integrated pest management using test kits and resistance breeding using marker technology or (in cases of durable resistance) genetic modification may reduce the need for pesticides with an overall positive effect on biodiversity.

The Overriding Socio-economic Context

In many countries, government policies and efforts of private breeding companies have stimulated the shift from food production for subsistence or for the local community in complex farming systems to production of fewer crops and breeds for the national or global market. Often this has implied a shift from many local, well adapted landraces to few modern high-external-input varieties. These policies and market efforts have resulted in a concomitant loss of diversity. These effects occurred in most developed countries but have now also become highly significant in most developing countries. Opening markets and globalizing economies, and the efforts of the international and national breeding institutes on major food crops are responsible for these effects. Higher food production based on high-external-input varieties and loss of traditional genetic diversity which we may need in future are two sides of the same coin.

Modern biotechnologies will strengthen these effects. Tissue culture techniques and artificial insemination require relatively low investment costs and limited expertise and are now well within the reach of most developing countries. Marker technology and genetic modification clearly require higher inputs, better facilities and more expertise, but are applied by several developing countries. And certainly, institutes which focus on agriculture in developing countries, e.g. the *Centro Internacional de Mejoramiento de Maiz y Trigo* (CIMMYT), the *International Rice Research Institute* (IRRI), the *Centro Internacional de la Papa* (CIP), the *Centro Internacional de Agricultura Tropical* (CIAT) and the *International Livestock Research Institute* (ILRI), have acquired these latter techniques and are involved in applications which will render novel breeding products for agricultural production. Only the *International Plant Genetic Resources Institute* (IPGRI) and the *Centres' System Wide Initiative on Plant Genetic Resources* of the *Consultative Group on International Agricultural Research* (CGIAR) are unambiguously dedicated to the rescue of agro-biodiversity that has remained. Future breeding strategies within the CGIAR Centres will probably have a major impact on conservation and utilization of genetic diversity in farmers' fields in developing countries.

In the CBD, loss of biodiversity in general and agro-biodiversity in particular has been recognized as undesirable. Improvement of production under low-external-input conditions, long-term sustainability, and future versatility of agricultural production are all dependent on a wide and rational utilization of agro-biodiversity. In modern agriculture, agro-biodiversity has received too little attention. Genetic diversity was regarded as being functional in breeding programmes, but not in the field where conditions are controlled and where uniformity is required for high yields. Development and application of technology in agriculture took place from this perspective, and largely ignored negative effects on agro-biodiversity. Conservation of biodiversity for the food security of future generations and for a more sustainable agricultural production requires a change in this attitude. It also requires a change in socio-economic conditions in agriculture which have caused the neglect for maintenance of biodiversity. This change should be based on the acknowledgement of the value of in situ conservation of genetic diversity and the need to consciously develop and use technology in this direction. This change should form a challenge for both governments and industry when opening new markets. Only then will the net effect of biotechnology on biodiversity be positive.

Agro-biodiversity and Seed Security: Some Considerations

Agro-biodiversity

The biological diversity related to agriculture is termed as 'agro-biodiversity'. It relates to the diversity among the agro-ecosystems, agriculture related plants, animals, avian, insects, microbes and the genetically modified organisms. Agro-biodiversity has remained the torch bearer of the humankind. It saved humanity from several catastrophes, ensuring 'something or the other' during such situations due to the underlying working principles, such as, endurance, avoidance, phenotypic plasticity, horizontal resistance, endemicity, etc. This particular recognition of agro-biodiversity requires several considerations, their factor analyses and looking for solutions. The key policy issues include determination of appropriate research and development (R&D) approaches, making developmental plans, and aiming for sustainability in the global, regional and national contexts. The agro-biodiversity could be viewed from two contrasting angles, namely, (i) the traditional agro-eco-systems, and (ii) the intensive farming systems.

Whereas the diversity among the cultivated plant species is maintained, more or less in equilibrium in the former case, it has to be guided externally in case of the latter. It is estimated that 20 percent of world's current food production is met from traditional multi-cropping systems which also help in maintaining ecological equilibrium and sustaining crop genepool admixtures, transgression and other micro-evolutionary processes.

The intensive cropping systems or the modern agro-eco-systems, on the other side, have evolved more recently, basically to enhance production *vis a vis* farm economic returns. A remarkable success has been shown in India harnessing the natural resources coupled with development and utilization of improved technology. It included a 275 per cent increase in the foodgrains production in the last fifty years, which resulted in a buffer stock of 36 million tonnes in 1994-95 and a record harvest of 199 million tonnes in 1996-97 as against a total production of only 51 million tonnes in 1950-51. This achievement was made in spite of a growing population pressure at 2.1 percent per annum.

The estimates showed that a meagre 8 per cent increase in food production has been achieved due to increase in area under cultivation whereas the rest 92 percent increase resulted from higher yield per hectare which has apparently been contributed by improved varieties, better agronomic packages of cultivation practices and better disease and pest control measures. The selection of improved types in all these cases could be made possible due to the availability of diversity along with expertise to select the right 'types'.

Estimates also showed that, on an average, more than half of the cultivated areas under major foodgrain crops, viz., wheat, rice, sorghum, maize and pulses were under high yielding varieties. Production in these systems is more dependent on the responsiveness of improved cultivars to various inputs, farm mechanisation, etc. and the various crops and varieties have coexisted albeit based on relatively limited research results. Several threats to agro-biodiversity have recently emerged in the advanced agricultural systems, particularly in many technologically advanced countries which posed several questions. Some of the relevant issues that emerged include:

i. safe conservation of landraces/farmers consciously selected heritage seeds and their management under a sustained system providing for further opportunities for agro-evolutionary processes; and

ii. an efficient, judicious and equitable use of this diversity to attain progressive increases in yielding ability and desired adaptability of improved varieties.

In this entire chain of events, seed of quality holds the key to ensure other agro-production inputs to be far more productive and cost effective. One can not think of an effective food and nutritional security unless the seed, being the basic input to crop oriented production, is assured.

Sustaining Agro-biodiversity

Institutional Mechanism Provided by the Convention on Biological Diversity (CBD)

Nations were required to evolve a *sui generis* system for varietal protection, go in for patenting or adopt a combination of both, but not in isolation from the broader goals set out in agenda 21 of the CBD. Such a *sui generis* system, in absolute sense, provides an institutional mechanism and legislative protection to manage and enrich *inter alia* the agro-biodiversity, within the national territory.

It would give due opportunities to on-farm conservation for sustainable use besides protecting and safeguarding the interests of those who conserve and maintain the vital components. Issues that the nations must consider, as per the requirements of CBD, include to plan and implement in an effective legal and/or regulatory frame work which would ensure no harmful effects even on territory beyond national borders upon consequent use of various components of agro-biodiversity; the regulation of exchange of germplasm is done in conformity with the charter of the UN and the international law, and that an intellectual property protection is provided to plants or plant parts in other member countries, etc.

Nations may be required to maintain two contrasting agro-ecosystems. First, the traditional agro-ecosystems encompassing and protecting biological diversity therein along with its biotic, abiotic and environmental components providing/maintaining due opportunities to sustainability and the micro-evolutionary processes.

The cost of maintaining such traditional agro-ecosystems, through imparting awareness and providing suitable compensations to the local folk who would practise it, is bound to be high but looking into the indirect impacts of such conservative mechanisms particularly on the climatic equilibrium, the CBD appealed to Global Environmental

Fund (GEF), in one of the decisions made in the third meeting of the Conference of Parties (COP-III), to make liberal financial grants for the above purposes pending settlement of the issue of establishing a Global Biodiversity Fund, which is being strongly advocated by several partners and fora. Lessons could also be learnt from the more recent indigenous knowledge and social framework of traditional monospecific cropping systems so as to help maintain the ecological equilibrium in advance, highly productive, intensive agro-ecosystems. Funding provisions may specifically be made for conducting research on these aspects.

Sustaining Alternate Ecosystems

A concern has been shown that the traditional agro-ecosystems should be conserved along with diverse evolutionary forms of cultivated plant species as well as farmers' conscious selections, particularly in the areas of origin and diversity of crop plant species and their relatives. This, obviously calls for concerted efforts towards developing a better understanding in terms of the following:

i. classification, indexing and inventorisation of traditional agricultural systems, still in vogue, with respect to their ecosystems, crop combinations, seasons and/or systems of cultivation;

ii. potential core area(s) under each, respective system or sub-system; and

iii. the custodian community.

At the same time, guiding and maintaining diversity in the advance agro-ecosystems through research, extension and legislation is considered essential to tackle several problem situations, such as, biotic virulences, abiotic stresses and adverse climates.

Rapid appraisal systems are, therefore, essential so as to maintain and guide requisite diversity in these agricultural systems to enable fast detection and appropriate replacement of the varietal components, such as those succumbed to diseases/pests/abiotic stresses or those which have been surpassed by more productive types in research.

The creation and maintenance of agro-biodiversity could be ensured through multi-crop, multi-varietal mosaic cropping which shall mimic the co-existence of cultivated plant species and the within-species diversity in the traditional agro-ecosystems. Also, the maintenance of the latter through awareness generation and providing suitable compensation to the local people who practise it, would be essential for an effective management of agro-biodiversity.

In Situ *on-farm Conservation*

There is a demanding need to enumerate agreed scientific principles and practices for 'on-farm' genetic resource conservation. It includes usable classification and nomenclature systems for landraces, as well as methodology for their characterisation. Due to lack of available information of this type, it is imperative that studies should be carried out on these aspects side-by-side while implementing 'on-farm' conservation programmes. A better understanding of the temporal and spatial movement of genes and varieties into and out of cropping systems and the causes of loss or recession of some varieties or traits over others should be highly useful in further setting the pace and direction of crop evolution. Adequate studies of some of the traditional practices and customs have not been carried out in most of the agro-biodiversity rich areas. Systematic surveys and collection of data on the traditional knowledge base, including methods to maintain healthy populations and seed, require an inquisitive follow-up and indexing.

The 'on-farm' conservation is generally associated with traditional agro-ecosystems and small scale farming although the farmers' participatory approach has been more advocated to help select suitable high yielding varieties (HYVs) for the advance cultivation systems. The large-scale commercial farms also have options to promote agro-biological diversity through crop rotations, intercropping, cover crops, green manures, paddy-cum-fish culture, apiculture, etc. The seed requirements for such situations would depend upon the adaptability, preference and acceptability of component HYVs for each sub-situation. Such practices need to be standardised as per diverse agro-ecological zones, so that sustainability of seed component is ensured under a dynamic intensive cultivation system. The more likely negative impact on sustainability of seed under the intensive cropping system may occur due to successively superior varieties being evolved at frequent intervals, the imminent displacement of modified agro-ecosystems which were created by cultivating such superior HYVs, the evolution of virulences or other environmental factors, such as, off-site pollution or accumulation of certain harmful gases like methane or ozone, etc. which need to be addressed by well planned, self-generating mechanisms.

Domestication of individual crop plant species is unique to agro-ecological sub-regions and localities and corresponds to cultural and ethnic diversity, including food habits, habitats, climates and available crop plant species. There is a purposeful need to maintain their representative seed 'on farm' because re-building new varieties for similar situations with same amount of adaptability and phenotypic plasticity would require strenuous planning, breeding efforts and resources, including funds and time. Capacity building for 'on-farm' conservation of traditional 'types' requires formulation of suitable policies, sectoral and cross-sectoral plans and programmes for implementation and also interim financial mechanisms.

Recognition of the associated local knowledge, innovations and informal practices would be needed to promote traditional farming systems, whereas establishing community seed banks in cooperative or private sectors, developing and maintaining cultivar/landrace registration system, etc., is required for smooth replenishment. The recognition of 'Farmers' Rights' is important along with formulation of implementable laws and regulations so as to ensure availability of the much needed 'seed' with due compensation or assurance of benefit sharing to its holders. Country level implementation of Registration of Plant Germplasm has been implemented in India by the ICAR and the planning/action related to other aspects is also initiated.

Compiling case studies for conservative, low external output based production systems through incentives for developing environmentally sound technologies for conservation, replenishment and sustainable use of biological diversity are issues which lack direction and, therefore, need sincere input from all concerned under the *sui generis* situations.

Geo-sectoral Approach

Contiguity of adjoining areas across the geo-political barriers, representing similar genetic diversity for particular species genepool, can be attributed as a basis to conserve, utilise and share the benefits of the *in situ*-on-farm conservation. Safeguarding the regional 'gene' interests could be ensured through sub-regional consortia and regional networking. The increasing role of Global Forum and the regional fora in strengthening relationships among the national agricultural research systems (NARS) and between various NARS and the Commonwealth Group of International Agricultural Research Institutions (CGIAR), could be viewed as new initiatives in desired direction which need to be pursued. Technological and technical needs can be gradually fulfilled for innovating, standardising and researching into the rapid appraisal systems for agro-biodiversity management using and strengthening the information network(s), as well as database system(s).

Due to the recent global interest for germplasm exchange and concern for protection of Plant Breeders'/Farmers' Rights there is urgency that the countries develop a regional consensus by way of the following:

i. by entering into agreement for undertaking joint explorations in the region for collecting genepools of crops for which the region is centre of origin;

ii. by adhering to effective plant quarantine so as to maintain an effective mechanism for exchange of risk-free germplasm;

iii. by developing *sui generis* methods of protecting the breeders' and farmers' rights by each country considering its own unique situation such that the benefits accruing thereof are also available to the other nations of the region; and

iv. by instituting funding mechanism and ensuring financial support for the regional PGR activities. It is pertinent to emphasise that the seed industry be facilitated and the security of certain vital components ensured.

Consequent upon the adoption of CBD and also implementation/ enforcement of General Agreement on Tariffs and Trade (GATT)/ World Trade Organisation (WTO) provisions, new issues and challenges of global significance have opened up. Accordingly, the sovereign rights of the nations on their bio-resources have been specifically recognised. As a consequence, it has become responsibility of the nations to conserve these resources and ensure their availability for their systematic utilisation on sustainable basis. Coupled with this, an equitable sharing of benefits accruing from the use of plant genetic resources (PGR) is an important issue that influences the other vital issue of access to PGR. The funding support from GEF has yet not set its pace which emphasises the need to set up a Global Fund which could *inter alia* help shape and refine the action points on seed security and the seed industry.

There is an urgency to address the identified areas in harmony with the other institutional mechanisms in the fray. Recognising that the gene rich countries have a potential to release naturally evolved variability suited to several diverse situations, there is an adequate need to match with a generous global commitment towards the funding requirement to sustain the traditional/naturalised diversity on-farm at certain representative sites located within the diversity-rich areas of the world. Some activities might include, but not limited to:

i. Regional exploration teams comprising experts from countries involved could be assigned survey, collection and bio-prospecting of agro-biodiversity. Whereas the hitherto unexplored bio-resources should be collected, evaluated, documented and shared for immediate use besides long-term conservation in the gene bank, the well adapted and popular cultures should be identified along with their companion crops/varieties and the cropping systems, for their on-farm conservation;

ii. Seed multiplication, testing and distribution programmes should be launched for the traditional varieties/folk cultures/ landraces which have edge over other co-cultivated variable forms and the farmers provided with critical inputs, either as subsidised or free-of-cost;

iii. Farmers should be encouraged to plant 'rest of the traditional cultures' in a small block of their cultivation field by ensuring seed of all the diverse component cultures from the agency/ community seed bank. The produce should be re-purchased at the premium rate of seed rather than commodity;

iv. In order to develop regional capabilities/capacities, the human resource orientation and development needs to be given top priority, giving equal weight to educating both genders;

v. Existing gene bank facilities should be utilised for long term conservation of seed representative of diverse types for posterity. The satellite, medium term facilities, on the other side, should be used to maintain active forms with a view to replenish the lost or diminished types in a catastrophe/crisis; and

vi. Exchange of germplasm seed needs to be regulated by strict phytosanitary regulations and quarantine measures so as to minimise risk from new pathogens/pests prevailing outside the region. Joint ventures in bioprospecting, biosafety and biotechnology, including cryopreservation, *in vitro* conservation or handling of transgenics/genetically modified organisms (GMOs) need to dwell upon principles of sustainability and synergies/mutual cooperation. All such activities require adequate funding support which should be catalysed and facilitated as per need of the hour.

The damage being done to land, water, flora, fauna and the atmosphere altogether poses the challenge of future food security. It is the human interference that created the crop plants from the wild to their present day commercial forms. Their agro-biological features and dispersal mechanisms have changed drastically since then. The commercial forms are absolutely dependent on human beings for their perpetuation and survival. It is, therefore, appropriate to look into the interests of the seed producers of the required/desired types. The management approach would require monitoring of the varietal component at frequent intervals so that the balance could be maintained between the interests of the seed producers and the phyto-ecological requirements of the varietal components. Involvement of both governmental and non-governmental bodies and the individuals, as also the genetic resource and patent literacy, constitute an essential elements of systematic sustainability and developmental plans.

Realisation in India and Efforts on PGR Conservation

It is recognised that a balance between biodiversity conservation and economic growth continues to be an issue of concern. In this connection, the conservation and effective management of plant genetic resources is essential for achieving sustainable gains in productivity. Accordingly, there is a need to formulate and implement an Action Plan at the national and regional levels in conformity with the situations and provisions of the Global Plan of Action. It is realised that the conflict of minds is universal. PGR were earlier held in trust as the 'common heritage of mankind' whereas recently the recognition of 'sovereign rights of nations over their biological resources' has come into effect. Accordingly, it is believed that the handling of issues related to seed security, distribution or replenishment in the areas of crisis needs more judicious treatment in conjunction with the agro-biodiversity related issues.

Agriculture at large, being the dominant sector contributing 31 per cent to gross domestic product (GDP) genuinely needs funding support for research, capacity building, infrastructure development, subsidy and grants, particularly in terms of attempting to maintain a guided composition of seed component over time and space. It has to simultaneously tackle the two contrasting dimensions of sustainability and development. Hence, a balanced, harmonised and optimised model of growth needs to be pursued.

The institutional mechanism provided globally by the CBD well recognised the need to harmonise with the International Undertaking on Plant Genetic Resources and its revision since 1993, in conjunction with the transformation in the purview of FAO Commission now redesignated as Commission on Genetic Resources for Food and Agriculture (CGRFA). The FAO's World Information on Early Warning System (WIEWS) clearly intends to safeguard the interests of agriculture through prospecting, country reports, replenishment of vital seed component in case of crisis, etc. in larger interest. It is felt that agro-biodiversity be treated as a specialised discipline as a force to reckon with. The Conference of Parties of the Convention on Biological Diversity and the CG Centres/Global Forum have all shown concern that agro-biodiversity needs to be recognised specifically and should not be treated generally. Hence, the actions/issues, problems, potentials, opportunities, threats, etc., which needs attention could be listed as follows:

- The issue of growth is central to sustain the farm, family and livestock. This would essentially call for a balanced approach in conversion, use, cycle and recycle.
- As the conservation and effective management of plant genetic resources is essential for achieving sustainable gains in productivity, an action plan at the national level is being drawn to give due emphasis on this component of biodiversity. The plan would be put to effective implementation during 1998.
- The tasks of PGR management would essentially have to be dealt with from two angles. The one in conjunction with other activities, and the other in collaboration with various stakeholders.
- Documentation of traditional wealth and knowledge of the country is well recognised. Documentation of traditional knowledge on all fields of plant wealth particularly on medicinal plants, skills, folk medicinal formulations, etc., is contemplated under a formulated programme to be implemented through a mission mode approach.
- The wild relatives of crop plants are the hope of future man-guided evolution of crop plants for posterity. The wild relatives of crop plants in India amount to over 325 species, which is a significant number and represents 1/5th of the total wild relatives' diversity available globally. There is a need to further study and exploit this rich treasure in the wilderness.
- A lot of national input has been made in raising a long-term collection of above 1.63 lakh seed samples of various crop plant

species and their relatives in the National Gene Bank. This facility as a National Heritage is supported from the national resource for its sustainability and strengthening in order to ensure conservation for posterity.

- Other modes of conservation are also being supported on case-to-case basis. The tissue culture repository with a holding of nearly 1,000 cultures of about 30 different species together with the cryo-preservation facility holding over 2,500 accessions of recalcitrant seed species on experimental basis, constitute part of the national gene bank. Field gene banks at different NBPGR stations have live trees of various horticultural plants, energy plantation and other economic tree species. Conservation of seed samples in the cold tracts at zero energy input is contemplated to supplement the collection at the National Gene Bank Repository. Research scale on-farm conservation activities are also envisaged and being pursued.
- Role of farmers in conservation of genetic diversity is well recognised. Farmers have to be convinced to practise on-farm conservation of their traditional cropping structures and system. This would give the world a hope to sustain and in getting naturally evolved forms in particular environments. For effective *in situ* on-farm conservation of genetic diversity, specific regions and species exhibiting concentration of diverse forms are being identified. It is believed that a system needs to be developed to find out the farmers' views and their expectations in sustaining *in situ* on-farm conservation of genetic diversity. Also suitable modes and mechanisms of incentives to farmers need to be developed so that the farmers resort to an on-farm conservation of the traditional types. Of the several options suggested, this may be done by expanding the Public Distribution System (PDS) to crops other than wheat and rice and also by chalking out market mechanisms for the traditional varieties of wheat and rice. Consideration should also be made to re-define productivity, so that not only yield but other components, such as, fodder (biomass) production, soil nutrient status, response to inputs, etc., are also accounted for and farmers get their due returns.
- To ensure sustained *in situ* conservation of agro-biodiversity, the areas/zones under traditional crops and continued traditional practices need to be earmarked and declared sensitive *in situ* on-farm conservation areas.

- There is need to develop instruments and provisions for Plant Variety Protection (Breeders' Rights), Farmers' Rights, Community Rights etc., in conjunction with effective linkages with the public sector, private sector, NGOs, Farmers' Associations and Communities for PGR conservation, in general, and collection and evaluation of germplasm, in particular.
- Emphasis has to be laid on human resource development to build up expertise in agro-biodiversity management, including legal implications and forms of access and benefit sharing. Germplasm collection is an expertise that would need stringent test over time when the depletion of resources further takes place globally. Germplasm management and conservation need equal priority. In this endeavour, the National Agricultural Technology Project is addressing the issue in right earnest.
- Recognising that information is a powerful tool and communication is the best means to generate awareness and to set the right track, India has launched a massive programme to address this issue in totality. Each district is provided on-line connectivity through electronic media.
- Preparation and dissemination of popular literature on various PGR related developments at the national and the international levels, for creating awareness widely, including at the School, College and University levels is envisaged. Updating the information on intellectual property rights (IPR), CBD, PGR Acts and the development of software for literacy are considered necessary with adequate budget.
- Access to PGR among the contracting parties is a matter that is still receiving international debate. Benefit sharing from germplasm collection already existing in the CGIAR Institutes collected from developing nations is also being considered at length. The issue is under debate to reach to a logical conclusion.
- Legislative measures, such as follows, are being taken up at the national level to cover the interest of intellectual property protection:

i. Plant Variety Protection and Farmers' Act (almost at final stage);

ii. National Biodiversity Act (at drafting stage); and

iii. Patent Act (Revised) (being debated in the Rajya Sahba through a sub-committee of Parliament).

- A national level Advisory Committee on Genetic Resources is in the offing to give a sense of direction on all related issues.
- Memoranda of Understanding (MOU) and Material Transfer Agreements (MTA) need definite input to face all types of challenges. Conflicting issues related to PGR have to be taken up and harmonised in several brain storming sessions. A few such deliberations are on.
- A sound database on genetic wealth is a pre-requisite for an effective conservation strategy. A huge expenditure is estimated for the inventorisation and genetic resources conservation at the national level which has to be generated mainly from public sector. A National Fund is contemplated.
- Among the issues related to conservation of agro-biodiversity, we must give a national treatment and priority to the seed gene bank at NBPGR and also include, besides the identification of conservation sites, the conservation partners. We must decide upon as to which sites for *in situ* on-farm conservation to begin with. To identify genuine NGOs and other partners for their involvement in PGR Conservation is an important task. Feed back information from the NGOs and other stakeholders is essentially required as well as the dialogue with seed companies and other private sector people engaged in PGR activities. We should decide on the above issues genuinely and initiate action.
- Open attitude, dialogues and transparency are important for the perpetuity of the national efforts. A need for substantial financial support is felt.
- It has been planned to equip all the co-operating centres/data generating sites (NAGS) with microcomputers and necessary software. The National Information System has to cater the needs of all partners of Indian National Plant Genetic Resources System (IN-PGRS) including other countries. A thorough system analysis has been done in this respect in terms of input and output including their generation and standardisation. Due consideration is also given to information flow, security/ protection against misuse, hardware and software, manpower and communication facilities. Efforts for development of on-line system, availability of the system on the Internet are also under active consideration under the aegis of ICAR's Agricultural Research Information System (ARIS).

Efforts at the Fao Level

The salient points related to the progress of the global system on conservation and sustainable use of plant genetic resources for food and agriculture which emerged in the seventh regular session of the FAO Commission on Genetic Resources for Food and Agriculture (CGRFA) at Rome during 15-23 May, 1997 may be reviewed as follows:

- A multi-year work programme on agricultural biodiversity established as per decision III/11 of COP shall be pursued by the FAO Commission and CBD Secretariat though cooperation. Similarly, FAO's monitoring/reporting on agricultural biodiversity shall be consistent with CBD, other relevant inter-governmental bodies and world food summit' Plan of Action.
- CBD through its decision II/5 has requested GEF to give priority support to efforts for conservation and sustainable use of agro-biodiversity. Many countries stressed the need to have separate funding! FAO commission requested FAO to provide assistance and guidance to countries on request in implementation/ monitoring of the Global Plan of Action (GPA). Further, the monitoring of GPA and updating/reporting on the state of world's PGR were held complementary activities which do not need separate reports (i.e., need only one report each) by countries.
- No changes in the CCGCT (Code of Conduct for Germplasm Collecting and Transfer) or any further work on Code of Conduct for Biotechnology were recommended, till the revision of IUPGR. However, a decision guide on regeneration of accessions in seed collections, prepared by IPGRI, has been made available to the FAO commission for comments by members.
- The existing arrangements between FAO and the 12 CGIAR centres shall be extended, pending the revision of IUPGR.
- Statutes of the inter-governmental technical working group on AGR (Animal Genetic Resources) and PGR were presented. Both working groups have 5 members each for Asia. The AGR were specifically addressed in view of the broadened mandate of the Commission and to facilitate integrated approach to agro-biodiversity.
- The summary statement presented by IPGRI informed the Commission that 0.5 million accession were now covered under the FAO-CGIAR agreements with the designation of 50,000 new accession under this arrangement.

- The summary statement presented by the CBD Secretariat showed that at least three decisions of the COP-III, Buenos Aires, Nov., 1996, were of direct relevance to the ongoing negotiation for revision of IUPGR, namely, Decision III/11-Conservation and Sustainable use of agro-biodiversity, Decision III/15-Access to Genetic Resources, and Decision III/17-Intellectual Property Rights. The COP-III noted the narrow options for legal status of revised International Undertaking, as being, i) a voluntary agreement, ii) a binding instrument, and iii) protocol to CBD. It encouraged parties to implement GPA at national levels. Pollinators and soil-micrograms in agriculture needed special attention (as bio-indicators) and comprehensive reporting by countries to Subsidiary Body on Scientific, Technical and Technological Advice (SBSTTA)/ Conference of Parties (COP) of the CBD.
- An identified group of competent people needs to resolve the PGR issues and help prioritise various PGR activities. There is need to identify crops/regions on priority.
- The dialogue between the public and private sector should continue and there is a need to create awareness about PGR.
- An information bank should be established to enhance literacy, management and utilisation of genetic wealth. Emphasis should also be given to regional languages in developing the information base.

World Information and Early Warning System

The FAO's initiative to develop a systematic information base on PGR and related technologies under the World Information and Early Warning System on Plant Genetic Resources (WIEWS) is aimed to collect, collate, maintain and disseminate facts and figures on country profiles, national PGR programmes, *ex situ* collections, database on PGR, existing crop varieties, seed production and supplying agencies, *in situ* on farm conservation, bioprospecting and exchange of seed.

The nature of such an activity requires a comprehensive information input and a grid approach for analysis and interpretation of data. The system is open and responsive to queries from all stakeholders and interest groups. In practical terms, it would be relatively simpler to attend to details of the resource availability and the custodian(s) but, considering the importance of implementing CBD and WTO provisions and the imminent revision of the

International Undertaking on Plant Genetic Resources, the access to resources for the purpose of replenishment could pose certain difficulties. The access would have to be required from any one of the following sources:

i. the *ex situ* seed collections;

ii. the seed multiplication and supplying agencies; and

iii. the local communities/community gene banks.

The implications, in each case, could be several and manifold. The access from *ex situ* CG system collections is being debated wherein two clear-cut options have been put forth.

The pre-CBD collections are still available freely which means their replenishment would be only a matter of time actually required for the multiplication of their seed. However, in case of requirement of post CBD collections, several access-related issues shall have to be sorted prior to the multiplication of seed for replenishment. The requirement of seed from commercial/public seed multiplication and supplying agencies could be easily met but such a replenishment would not serve the purpose effectively. This being so, the agencies could provide seed of few dominant varieties, which they are currently handling, in contrast with the requirement of a wider genetic base.

The replenishment of seed from farmers' varieties/land races faces two different situations:

i. the issue of consent/access; and

ii. the issue of compensation or sharing of benefit accrued from the use of their heritage seed.

The setting up of community seed banks and the Global Biodiversity Fund have definitely been provided as suitable institutions which could help in build up at an appropriate system and come to the rescue in cases of crisis. The setting up of a separate Global Biodiversity Fund is still being debated as also the Farmers' Rights.

The development of the *sui generis* systems to meet the countries' own (local) needs, has to be accelerated. Through this mechanism, the whole idea is to encourage local level replenishment through procurement of required seed from nearly/similar climate logical areas.

Some Queries Seeking Answers

Increased investments on *ex situ* conservation are unavoidable with each increment in the quantum of work and management skills.

The cost/benefit ratio does not work in such cases because the conservation is for posterity. In this endeavour, the obvious questions seeking answers would be:

i. What should be the estimate of funds for genetic resources conservation at the CG Centres, at the National genebanks and at community seed banks?
ii. What kind of activities we need to take up in the beginning which ensure fast replenishment in cases of crisis?
iii. What shall be the mechanism of funding-whether the Global Biodiversity Fund or private sector contribution shall be forthcoming or not?
iv. Which sites for *in situ*/on-farm conservation to begin with?
v. How to identify genuine NGOs for their involvement in PGR Conservation?
vi. What could be a standard format of collecting and collating information for WIEWS on PGR *ex situ* conservation, country status, on-farm conservation and wild relatives? Do the existing proforma suffice?
vii. What would be a suitable indicator or an appropriate method for monitoring of genetic erosion for use under Early Warning Systems?
viii. What should be the suitable proforma for multicrop passport data?
ix. How to generate compatible country reports providing information for the State of the World's PGR and the action on GPA?
x. How to minimise diversity and redundancy in use of computer software for various Database Management Activities?

Policy on Seed Security for Effective Intervention

Primarily there are two broad issues:

i. Availability of seed for bringing much needed improvement to meet the existing and emerging needs of uninterrupted varietal improvement; and
ii. To have in hand seed for direct intervention and replenishment/ utilisation for commercial cultivation to meet the normal and or contingent situations.

On the first issue, collection, conservation, maintenance, documentation, storage, mode, mechanism and system development for facilitated access with pros and cons on a balance sheet need to be put forth to assess various policy implications for exploitation of genetic resources on a long term sustainable basis.

To deal with the second issue, concerted efforts on a long term perspective plan basis would also be urgently called for as follows:

1. As we go along to harness the productivity on a higher plane with the stipulation to attain the maxima; and the maxima being always dynamic in space and time; and with the localised increase in biotic and abiotic pressure in many of the location specific zones, tailor-made varieties to harness the potential of the eco-region and the system would essentially be called for. It would ultimately be the system's productivity on a balance-sheet which would provide enhanced productivity and profitability by balancing the input and output and the final turnover. In this endeavour, there would be a 'contradiction' as potentially the entire crop improvement programme rallies round conversion of solar energy by means of a biological system which is portable and marketable, be it food, feed, fodder, fuel or fossil. In order to attain efficiency of a high order of a plant system, the genetic tailoring at a narrow plane would obviously be the most crucial factor. This would hinder wide adaptability and thereby put a problem on stored seed utility to meet the contingent situation. If one needed to implement a seed security plan of a localised nature to meet the specific needs to harness higher potential, it would not be without a cost and obviously the cost would be of a high order as compared to the one in which a variety of a wider adaptability is considered. The cost would be lower in storing the seed of a widely adaptable variety as the manoeuvring and likely utility of the seed would be of a high order and, hence, the stock would be renewable and utilisable in a wider way.
2. To ensure that the tailor-made varieties are available, location specific breeding efforts would essentially be called for. At the same time, a finished product would need to be tested in different regions, situations and systems, be it of public or private origin. If of private origin, in particular, an instrument of plant variety protection would be imperative so that first rate technology at the global level could flow for the benefit of the farmers.

3. Provision of registration of varieties, system of varietal testing for their wider availability and suitability for cultivation cutting across countries and their documentation would be called for so that first hand knowledge is available as to what is suitable and for where?
4. A consensus and concerted efforts would be imperative for buffer stocking of seed. As a calculated risk and based on past experiences of the kind of calamities and eventualities, buffer stocking of seed to meet the national needs could be thought of. This would be an ex-ante decision as to what would be the kind of varieties and to what kind of situations so that only their seed is produced, stored and replenished. The private sector, keeping in view the market needs and likely commercial venture could also consider seed production and its storage and in turn need based marketing. It is quite likely that situation would demand for seed stocks to meet national calamities as has been witnessed in certain parts of the world, say in Ethiopia, where a good amount of seed was rushed from neighbouring countries. At times, varieties were suitable and many times they were not at all congenial for commercial cultivation, but with little option.
5. It is a recognised fact that pest and pathogen move with the seed and in the liberalised world the problem of phytosanitary maintenance and post entry quarantine would be far more than ever before. As such, these regularity facilities would essentially be required to be strengthened particularly in case of seed borne diseases and viruses. It would be necessary that, in general, a common code of conduct is formulated and implemented after international consensus building. FAO can play a prominent role in this direction. Seed certification to ensure the quality of the seed would be of paramount significance. Efforts should be made so that strict provisions are adopted and implemented by various nations so that their certificate is globally recognised and supply of quality seed is ensured.
6. In the present world, the issue of free or open market has been discussed considerably. But one needs to precisely understand that it is not a free market or an open market or a liberalised or a globalised market. In fact, it is a competitive market in terms of quality and cost and it is the quality to which the

global environment must suit so that the one which can bring added returns on investments made is allowed to flourish through seed culminating into food and feed security in the posterity.

7. Under the WTO, nations have agreed to enact a system of plant variety protection. The system must be put in place so that a congenial environment is provided for greater investment in plant breeding efforts and consequent seed production and its dissemination. The protection by appreciating the intellectual property and ensuring returns on the investments made to some of the existing problems on quality seed production and its availability could be overcome. Hence, intellectual property protection as a means for ensuring added investment be appreciated and much needed incentive and reward system be created commensurate with the dimensions of the problems. Lest, seed security would be difficult to be ensured and hence, food and nutritional security would be jeopardised.
8. For ensuring much needed nutritional security at the household level through varietal constellation, it would be desirable that quality aspects of the produce are appreciated and the worth of the produce is provided in the market which could trigger quality related improvement programmes in the right direction.
9. The estimates at the global level has clearly brought out that to ensure food security, productivity of foodgrains will have to be doubled in the next 50-55 years. This would happen only if accelerated efforts on varietal development front is made and much needed insulation through resistance breeding programmes is put in place for various biotic and abiotic stresses at the earliest possible. It is recognised that apart from the varietal potential *per se*, the cropping intensity will have to be essentially enhanced where net monetary returns per unit area, input and time should be the determining factor and system productivity in totality needs to be the determinant. This would call for a system wide system approach on varietal development, seed production, storage and its effective utilisation in a region, situation or system. The primitive land racers or cultivars are virtually a blend of types, closely related. Many times, such mixtures of types provide greater insulation in production because of their inherent varietal blend situation. This kind of scenario is particularly effective in terms of

moisture stress or salt stress or even in case of disease and pest problems where horizontal resistance is automatically brought in the fore front. There are very limited efforts on improving the potential of land races, although many of them could be of great consequence to meet effectively the local situations. In this endeavour, farmers' participatory approach to plant breeding is worth considering. But this approach could be adopted to regions, situations, crops and commodities where improved production technology has not been brought to bear on the productivity front due to improved varieties or monoculture practice of cultivation. The production of identified promising types, their storage of seed and its use can go a long way in environmentally disadvantageous situations, be it drought prone or hilly regions or coastal situations. Hence, the issue needs concerted thinking, pragmatic approach and well-drawn policies and programmes to improve the existing land races by mere selection and following the pedigree method of breeding involving local people who would be the end-users.

4

Prospects and Limitations of Agricultural Biotechnologies in Forestry

Introductory Remarks

It is now generally that the practice of plant breeding will move forward as progress is made in knowledge and technology. Breeders in the future will not only be forced to hone their traditional skills but will also have to integrate them with knowledge and experience to coincide with more recent advancements and technologies derived from cellular and molecular concepts and approaches. For worldwide crops such as potato, rice, maize and in the future wheat, as well as for high value vegetable crops, the future of breeders depends upon the use of a combination of breeding practices (traditional breeding, genetic engineering and tissue culture).

Thus most aspects of agriculturally relevant biotechnology are covered. Plant tissue culture techniques not only provide essential ways for the clonal propagation of many agriculturally important crops(e.g woody plants ornamentals and vegetables) but they are also the basis for the production of transgenic plants and are at the fore-front of recent studies in plant physiology, developmental biology and biochemistry. Spectacular progress in our knowledge of plant sciences, including those relevant to agriculture and breeding, can be expected from the integration of more traditional sciences with mole-cular and cellular concepts and techniques. A prime example is seen in the impact molecular genome analysis is already having on traditional breeding (via so called map-based breeding) but also on research and molecular breeding (e.g, via map based gene-cloning) This rosy description of the future of genetic engineering is richly tempered in this book by a realistic assessment of the many problems that still need to be faced before what is predicted can be accomplished. It was therefore wise to ask scientists well known for their "no-nonsence" approach such as Ingo Potrykus and his colleagues to write about genetic engineering of crop plants.

To quote from their introduction: " it has been possible to develop the state of the art of gene transfer to plants to a level at which many of the major crop plants are accessible to the technique and for which we have good reason to believe that there is no basic biological principle that will prevent gene transfer to a specific crop plant " Yet these authors clearly identify the considerable research and development that is still needed before one can, in many but not all instances, consider large scale applications.

A recurrent point made and well documented in various chapters of this book is that genetic engineering-because it is not limited by the natural barriers that ensure the maintenance of species and prevent transfer of genes between unrelated organisms-is an extremely powerful way to increase the genetic variability available to breeders. The present rise in interest to acquire new knowledge via fundamental research can best be illustrated by reminding ourselves of recent progress in our understanding of the molecular physiology of development, of the mechanism of action of old and newly discovered phytohotmones and of growth signals used by microorganisms that are pathogens or symbionts of plants (such as for instance lipo-chitooligosaccharides). Biotechnology can contribute to making traditional agriculture more productive while at the same time more sustainable. It can also help achieve the goals of "non-food" and or cash crops agriculture as has been described in the various chapters dealing with crop improvement and metabolite production. Hopefully sufficient attention will be paid to regional crops of importance by local populations in developing countries.

Finally topics such as "Biofertilization" and "Bioremediation" are convincingly dealt with Topics such as "Biotechnology of Farm Animals" and "Marine Biotechnology" as well as the unavoidable but all important "Legal and Public Aspects" round out the all inclusive title of Agricultural Biotechnologies. Some topics have already come to blossom and illustrate well the potential of present day plant sciences. There is the breakthrough in the cloning and characterization of resistance genes along with significantly improved understanding of the nature of avirulence genes and elicitors. Many hypothetical models predict that resistance genes should be involved in the signal transduction pathways linking pathogen derived signals with the regulated expression of defence genes and of programmed cell death (hypersensitivity reaction). The first available data on resistance genes show that this concept may well be correct in its very general sense, even if all resistance genes do not code for elicitor receptors.

We may safely predict that in the near future we will know more details of the molecular mechanisms underlying plant-pathogen resistance mechanisms. Several elicitor receptors will soon be isolated and characterized along with genes involved in localized pathogen-induced programmed cell death. Plant breeders will make use of isolated resistance genes to produce crops with improved resistance to a variety of pathogens and pests. The content of this book fully justifies the conclusion that Biotechnologies are bound to play an ever growing role in agriculture not only in the long term future, but also presently, and in the near future.

Fortunately the progress in applied plant biotechnology corresponds with and is in fact stimulating fundamental scientific progress. This is important because the real driving force behind this science is the success of applied plant biotechnology such as: 1) plant protection based on the expression of several different introduced genes coding for different principles acting on the same targets and thereby providing a possible solution to the problem of the emergence of resistant pathogens and pests, 2) new quality traits as well as future crops producing tailor-made non-food products (lipids, carbohydrates, and biodegradable thermoplastics). Only if plant biotechnology becomes a commercial as well as an environmental success will sufficient support be available for this fascinating research.

Plants and Agriculture

Human population has grown steadily since the advent of agriculture, nearly 12,000 years ago, which ensured a continuing and reliable supply of food. The most rapid growth occurred during the past two centuries during which time agriculture became highly industrialized, public health improved and there were increases in incomes and international trade in food grains. Despite the age-old human concerns about balancing population and food supplies and many population control measures the worlds population is projected to continue to grow into the early decades of the 21st century, stabilizing at 9-11 billion sometimes during 2030-2050. Much of this growth will occur in the already overpopulated underdeveloped and poorer regions of Africa, Asia and Latin America, which will be home to nearly 90% of the human population.

Food shortages were common during the early and middle years of the 20th century. The introduction of Green Revolution varieties of wheat, rice and maize during the mid-1960s reversed this trend and helped avoid major food shortages by keeping increases in food productivity slightly ahead of population growth. During this period, both China and India, the two most populous nations with chronic food shortages, became bet food exporters.

Realistically however such increases in food productivity can not be sustained indefinitely. It is not surprising, therefore that increases in food productivity have begun to decline during the past few years. Furthermore, improved economic conditions in China and India have created greater demand for better and varied food products, particularly poultry and meat, which require greater supplies of feed grains. As a result, both China and India have now become net importers of food. World food reserves have declined from a high of 77 days to less that 50 days. With the current trends of population growth and agricultural production the demand for food in the most populous parts of the World will double by the year 2025, and nearly triple by 2050, Increases in food productivity of this magnitude can not be brought about in such a short period of time by conventional breeding, especially when some of out most important crops are approaching the physiological limits of productivity, not by increasing the amount of arable land, which accounts for 97% of all food production in the world..

Arable land, which is finite and comprises about 3% of the earth's surface, is deteriorating and decreasing as a result of soil erosion, salinization, over cultivation, and acidification. These factors, combined with expected increases in population, will actually decrease the global per capita arable land from the current 0.28 hectare to 0.17 hectare by the year 2025. In addition, fresh water supplies, essential for modern high-input irrigation agriculture, are becoming limited by increased human and agricultural use, and polluted by agricultural run-off and widespread use of agrochemicals. It is feared that the resulting food shortages in the overpopulated parts of the world during the 21st century may lead to widespread social, economic, and political unrest, making food security the single-most serious threat to international peace and security. At the same time it is well known that countries with efficient agricultural systems generally have high living standards, strong economies, lower rates of population growth and democratic forms of government Increasing food productivity in a sustainable manner will, therefore, not only provide adequate nutrition to the expanding humanity, but will reduce population growth, protect the environment, promote economic development and ensure social and political stability.

The challenge for the agricultural sector during the next few decades is therefore clear double food production by 2025, and triple it by 2050, on less per capita land, with less water, under increasingly challenging environmental conditions. The situation is further complicated by the fact that in spite of the heavy use of agrochemicals, modern agriculture still loses nearly 42% of crop productivity to competition with weeds and to pests and pathogens, and an additional 10-30% to post-harvest losses due to a variety of factors, especially in the developing countries where storage conditions are poor. During the 20th century traditional plant breeding has brought about enormous increases in crop productivity.

However, plant improvement by hybri-dization is slow, and is restricted to a very small gene pool owing to natural barriers to crossability. Beginning in the early 1980s, advances in plant cell culture can genetic transformation have overcome these barriers by making it possible to transfer defined genes into all major food crops, including cereals, legumes, cassava, potato, and many vegetable and fruits. The entire global gene pool-whether it be plant, animal, bacterial or viral-is available for utilization. The first genes that have been integrated into crop species provide resistance to non-selective and environment friendly herbicides, and many pests and pathogens. Increasingly large acreages of transgenic maize, soybean, potato, tomato and cotton are being commercially grown for human use and consumption. Carefully planned introduction of such crops on a worldwide scale would greatly help in reducing or even elimination the enormous crop losses attributed to weeds, pests and pathogens. The use of such crops will also have a beneficial effect on the environment by significantly reducing the use of agrochemicals. Other genes for improving crop productivity, and manipulating starch/protein/oil quality and quantity, resistance to environmental stresses such as temperature and drought are also being isolated and studied. In the foreseeable future, these will be used to produce second generation transgenic crops. It is clear that during the next few decades a wide variety of transgenic crops will become integrated into agricultural systems in the industrialized countries, Introduction of such crops into the developing countries, which need them most, will be slow and largely through the efforts of multinational biotechnology companies, because most of the developing countries currently lack the scientific and industrial infrastructure to develop and introduce these technologies into their agriculture.

In the long term however, much of the increase in food production to meet the dual challenge of population growth and food demand must occur in the developing countries. This will impact greatly on their overall economic development, which will help control the relentless increase in the population. It is critical, therefore that scientific and technical manpower and infras-tructure be created in developing countries to take scientific and technical manpower and infrastructure be created in developing count-ries to take advantage of the remarkable advances in agricultural biotechnology.

Assistance should be provided to the developing countries be international organizations such as the United Nations Educational Scienticfic and Cultural Organization (UNESCO), the Food and Agriculture Organization of the United Nations (FAO) and the World Bank through its many international agricultural research centres, like those in Mexico and the Philippines, where Green Revolution originated, International agriculture in the latter half of the 20th century was dominated by the Green Revolution, Considering the power and potential of agricultural biotechnology, it is likely that the Gene Revolution will dominate the agriculture of the 21st century. Like the Green Revolution, it may help accelerate the rate of food productions, save lives from hunger, create livelihoods in rural households, save large tracts of land that would otherwise be needed for food production, reduce birth rates, ensure low food prices, stimulate broad-based economic growth, expand world trade, and help create a sustainable agricultural system for future generations. Finally, it should be understood that although plant breeding has long been, and will continue in the future, to be indispensable for plant improvement, it must now be complemented and supplemented by molecular breeding and genetic transformation, in order to establish a sustainable agricultural system for the 21st century.

Animal Biotechnology

Animal Biotechnology has developed rapidly from the early 1980s when the first transgenic mice and first in vitro produced bovine embryos occurred. Today animal breeding companies are using marker assisted selection to provide earlier and improved selection of breeding animals. Gene mapping efforts are annually identifying a large number of candidate genes for testing of usefulness in marker assisted selection.

The propagation of genetically valuable or transgenic animals is enhanced by in vitro production of embryos. There are already more than a dozen commercial companies selling or producing for customers in vitro produced embryos. Cloning of embryos of cattle and sheep has been shown to be possible, but low efficiencies have thus far prevented commercial use. Animal producing and genetic selection systems can be made more efficient by sexing the sperm used for in vitro production of embryos. The flow sorting of sperm cells to separate X from Y bearing cells has been successful in most species tested and in cattle and swine has resulted in offspring of the desired sex. While effective with in vitro embryos production, the sexed sperm are of less than normal viability and sperm numbers and not yet useful in artificial insemination or the freezing of sperm.

The artificial insemination and embryo transfer industries are today heavily dependent on use of cryopreserved sperm and embryos. However the cryopreservation of oocytes and the combination of some biotechnologies such as embryo biopsy and sexing with cryopreservation are still in the research stage. Transgenic cattle, sheep swine and poultry have been produced primarily by microinjections of DNA into pronuclei of eggs or by use of viral vectors. The intended use has been to change animal growth, disease resistance, wool production, or to produce new products in milk. The efficiencies thus far have been much lower than in mice. Therefore there has been little commercial use of transgenic technology except for the involvement of several companies in production of pharmaceutical products in milk.

The use of cultural embryonic stem cells and nuclear transfer to make offspring has resulted in calves and lambs. This technology offers promise of making large numbers of nearly identical offspring or more efficient gene transfer or selection. However thus far the technology is not sufficiently efficient for commercial use. Growth hormones and growth promoting factors are being developed for improved efficiency of animal growth. Growth hormones are commercially in use at present for promotion of milk production in cattle. The anatomy and physiology of birds and the late stage of egg laying prevents the introduction of DNA into pronuclei as in manuals. Thus other methods such as oocytle microinjection of DNA, viral introduction of DNA and introduction of DNA into primordial germ cells are being used in avian gene transfer research. Oocyte microinjection, blastoderm transfection and primordial germ cell manipulation have all been successful in producing transgenic birds. The production of transgenic birds appears possible but is not yet in commercial use. Patented applications of recombinant DNA biotechnology to animal agriculture became apparent in the early to mid 1980s with development of recombinant vaccines effective against calf scours and against rabies. Today numerous vaccines in use are produced by recombinant technology. Biotechnology has also revolutionized diagnostic techniques by introducing higher sensitivity and specificity. Monoclonal antibodies, DNA probes, Southern and Western blotting, DNA and RNA amplification through PCR, constitute as integral part of the diagnostic arsenal at the disposal of the advanced laboratories.

Based on these techniques rapid animal-side field test used by the clinicians furnish results in real time, permitting immediate decisions as to the proper treatment, vaccination, or stamping-out policy. High technology based diagnosis is expected to become a routine in every diagnostic laboratory forming a sound basis for control and treatment of animal diseases. In the short time of approximately fifteen years, animal biotec-hnology has moved from an unknown and unstudied science to a strong discipline of science and extensive commercial application. The science and its commercial use are rapidly progressing. The ultimate benefactor will be humans through improved food supply, nutrition and health care products available to humans around the world.

Biotechnology in Agriculture

During the last decade, tremendous progress has been made in the area of plant cell, tissue & organ culture. In vitro techniques constitute an important component of biotechnology, and have the potential not only to improve the existing cultivars, but also for the synthesis of novel plants & early release of high-yielding plants resistant to various diseases, pests, stresses & temperature. Green revolution of recent years has been the result of efforts that led to the improvement & better selection of species of agronomic interest. Some specialists feel that biotechnology is the second phase of green revolution, studies are already in progress to obtain crop plants having higher photosynthetic efficiency, improved fixation of atmospheric nitrogen, increased nutritional quality & greater tolerance of salinity, alkalinity, acidity, deficiency of nutrients & stresses. Wide hybridization through embryo & ovule culture both at the interspecific & intergeneric levels have been achieved in a number of crops through the culture of the hybrid ovules & embryos.

In vitro fertilization techniques have also been employed to overcome incompatibilty. Haploids produced through anther culture &

their diplodization have resulted in the release of varieties in crops such as wheat & rice.

Biotechnology

The term biotechnology is composed of two words-bio (Greek-bios-means life) and technology (Greek technologia-means systematic treatment). In other words, it is the science of applied biological processes. Accordingly to the oxford Dictionary, technology is the "Scientific study of the practical or industrial arts". Spinks (1980) defined biotechnology as, "the application of biological organisms, systems or processes to manufacturing & service industries".

European Federation of Biotechnology (1981) defined biotechnology as, "the integrated use of biochemistry, microbiology & engineering sciences in order to achieve technological (industrial)application of the capabilities of micro-organisms, cultured tissue cells & parts thereof".

In the report Biotechnology: A Development plan for Canada (1981) biotechnology is defined as "the utilization of a biological process, be it microbial, plant or animal cells, and their constituents, to provide goods & services".

Production of Pathogen-free Plants.

Plants traditionally being vegetatively propogated are systemically infected with viruses & other pathogens which greatly reduce yield & also Q uality of the marketed commodity. Virus diseases like potato leaf roll virus (PLRV) or potato virus Y (PVY) for example, can cause upto 95 % reductions in the tuber yield of potato crops.

Increase in yield upto 300% (Averaging 30%) has been reported following replacement of virus-infected stock with specific pathogen free plants. Since majority of viruses infect plants in a systemic manner their elimination may be achieved through meristem tip culture. Plant species for which virus-free plants have been obtained.

Species Virus Eliminated

1. Allium sativum (Liliaceae) Garlic mosaic virus
2. Brassica Oleracea (Brassicaceae) Cabbage black ringspot virus, Turnip mosiac virus, cauliflower mosiac virus
3. Chrysanthamum sp. (Asteraceae) Chlorotic mottle, Green Flower, stunt vein mottle, virus B.
4. Dahlia Sp. (Asteraceae) Dahlia mosaic virus, Tomato Aspermy, vein mottle, virus B.

5. Glycine max (Fabaceae) Soyabean mosaic virus. Impomoea batatas Feathery mottle, Hanmon Mosiac, (Convolvulaceae) ' Rugosa mosaic
7. Lilium Sp. (Liliaceae) Cucumber mosia virus
8. Monihot Sp. (Euphorbiaceae) African cassava mosiac
9. Musa sp. (Musaceae) Cucumber mosiac virus
10. Nicotiana tobacum (solanaceae) tobacco mosaic virus
11. solanum tuberosum(solanaceae) potato virus.

Although the apical meristems are often free of viruses, this cannot be regarded as a phenomenon of u niversal occurrence. The success of meristem tip culture depends upon several factors. One of the most important factors is the relative distribution of viruses in the growing tip of donor plants. There are some viruses which invade the growing tip viz, TMV, potato virus X, cucumber mosaic virus. In such cases virus free plants are obtained by combining meristem-tip culture with high temperature treatments.

Production of Disease Resistant Plants

This is one of the most useful applications of tissue culture in crop improvement. In potato, somaclones have been screened for both late & early blight resistance. In maize, somaclonal variation has induced resistance to race T of southern corn leaf blighrt. In sugarcane, resistance to diseases like fiji & downy mildew have been recovered. In lucerne, selection of cell lines & plants resistant to the toxin of *Fusarium oxysporium* has been accomplished.

Improvement of Nutritional Quality

One of the major sources of protein for human & animal commnsumption is constituted by the proteins contained in seeds of many plant species. The cereals & legumes which are major sources of storage seed proteins, contain limited amount of certain amino acids which are essential for human beings. Majority of these cereals are deficient in lysine whereas legumes are deficient in sulpher amino acids. A wide range of approaches have been employed for improving nutritional quality of various crop plants.

Important among them are selecting cell lines resistant to amino acid or their anologues of lysine, tryptophan, proline & phenylalanine.Isolation of variants over producing specific amino acids in culture has been successful, but expression in the whole plant & especially in the seed has not yet reached the level required to make an impact on protein quality.

Selection for Salt & Draught Tolerance

Continued efforts to increase intensity of cropping for increasing production from limited land resource by extending irrigation facilities have resulted in the gradual build-up of salt concentrations in the soil. This has resulted in loss of productivity on such soils. Salt tolerant lines have been produced in crop plants such as tobacco, tomato, cereals. Salt tolerance has been incorporated into rice lines with improved plant type. Some of the improved rice cultivars viz., IR 42, IR 43 & IR 52 are sált tolerant.

Development of cultivars tolerant draught can contribute significantly in agriculture economy. Tolerance to draught is a polygenic trait & involves highly complex osmo-regulatory functions. In tissue cultures, simulated draught conditions have been achieved through incorporation of non-penetrating osmotic solutes such as PEG (polyethylene glycol) and dextrans in the media.

Production of Genetically Variable Plants

The success of any crop improvement programme depends on the usable genetic variability in the base population cells in culture offer an excellent systems for inducing variations & regenerating pure mutant types. Genetic variation can be an option to lessen our reliance on cost intensive germ-plasm collection & conservation programmes. Variant producing capacity of cell culture can be augmented to a great extent by employing physical & chemical mutagens. Somaclonal variation has been extensively exploited for the improvement of a sexually propogated crops viz., potato, sugarcane. From cell cultures, some superior cultivars have already been produced in sugarcane which are high yielding, drought resistant & temperature tolerant.

Biofertilization

Molecular nitrogen in the atmosphere is converted into Biologically converted forms by nitrogen fixing micro-organisms e.g. Rhizobium. The most sophasticated approach to biofertilization is to create plants that possess the genetic capacity for nitrogen fixation. Attemps are being made to transfer genes for nitrogen fixation (hifgenes) from bacteria to plants.

Rapid Clonal Propagation

Tissue culture has found its best commercial application in production of cloned plants at a very high rate as compared to conventional methods. It is important specially for initially building up of propagation stock of elite clones or individual plants which are otherwise slow to multiply. A number of agriculturally important plants have been clonally multiplied.

Genmplasm Storage

The primitive cultivars & wild relatives of crop plants constitute a pool of genetic diversity which is invaluable for further breeding programmes. There are over 20,000 plant spices which are rare or threatened with degradation of their neutral habitats. The most economical form of storing germ plasm for seed propagated species is seeds. However, there are certain limitations of this method. Therefore, various methods of invitor storage of germplasm are thus of great practical significance for long term storage of germplasm. Presently, there are two approaches to invitor germ plasm storage. Slow growth technique & cryopreservation.

Biological Control of Agricultural Pests

Insects consume nearly one-third of human food supplies on earth. One of the major goals of biotechnology is to develop target specific biological pesticides that will kill some specific pest but will not harm other species. Several insect-specific pathogens, developed through biotechnological processes, are being produced commercially for their use as microbial pestices.

Bacillus thuringienis inocula have been used as bio-insecticides. Toxin produced by the bactenium kills the gypsy moth. The prevention of crown gall disease formation is achieved by spraying young plants or seeds avirulent species *Agrobacterium radiobactor* var, *radiobactor*. Microbial insecticides have a number of advantages over traditional chemical insecticides:

a) they are specific for a small number of species of insects & do not kill plants, animals & beneficial insects.
b) they are cheaper than organic pesticides and
c) they do not leave any residue effect.

Selection for Herbicide Resistance

Herbicides are used to control weeds in agricultural fields. Tissue culture techniques offer several advantages in developing herbicide-resistance plants:

1. The techniques for selection are relatively rapid;
2. Small volumes of cells represent the genetic potential of hectares of fields.
3. Metabolic studies are facilitated by uniform & sterile conditions &3
4. Possible new types of resistance not found in nature may be obtained from plant cell cultures.

Attempts to produce herbicide tolerance using somaclonal variation & invitor selection have already been made.

Biotechnology and Indian Agriculture: The Challenge of the Next Millennium

The world has won the important battle in the area of food security, but the war is still on. A total of 800 million people, that is one out of every six persons, in the developing world do not have access to food. One-third of all pre-school children in the developing countries are food insecure. We, thus, have a big challenge ahead of us as we enter the next millennium. It is true that the mass starvation that was predicted for Asia in seventies and eighties did not occur. It is only because Science was effectively put to work to raise agricultural productivity. In India, the ' *Green Revolution'* was a success due to the introduction of improved seeds, fertilizer, irrigation and plant-protection measures combined with positive policy support, liberal public funding for agricultural research and development, and dedicated work of farmers.Notwithstanding all-round achievements, the basic problems of food security, poverty, equity and sustainability, continue to be a cause of concern in India today.

The Indian Challenge

There are several concerns about Indian agriculture that we need to address. Even in the green revolution crops, rice and wheat, the

Indian average yield ranks around 50th in the world. A weak agro-based industry, the problems of weak marketing, storage, transportation, credit support, etc, haunt us. Only 0.3% of agricultural GDP is spent on agricultural R &D. The population is growing at an alarming rate of 1.8% per annum. It will be around 1.3 billion by 2020 and would require 50% additional foodgrains. A total of 200 million people are still below the poverty line and have insufficient access to food. Declining resource base, degeneration of natural resources, including soil fertility, water environment, etc, are some other concerns.

Food production systems centre around a few states only. Out of 17 big states, only 7 are self-sufficient and surplus in foodgrains. There was a significant decline in per capita intake of calories and protein between 1975 and 1990. Inspite of all these problems, there is plenty of good news. India harbours a vast diversity of basic natural resources. Tremendous bio-diversity of agriculturally important micro-organisms exists, which are not yet been harnessed. We have a variety of crops, including quality fruits, vegetables, foodgrains, and a number of plantation and commercial crops. Indian fisheries resource has spread over 8,100 km of coastline on the east and west, and provides access to marine resources and transportation.

Very large proportion of area of various food crops is under low productivity category. It can be increased with inputs of high class science. There is a scope to cultivate sizeable portion of waste lands through soil amendment and introduction of agroforestry. Several million hectares of saturated soil can be potentially exploited from rainfed lowlands of eastern India.

The India Exclusive Economic Zone (EEZ) of the marine sector, of about 2 million sq.km., has high potential harvestable yields. Only about 20% of the cultivated area is covered under tractor cultivation, great scope of making quantism quantum jump in production and productivity exists, changing consumption patterns offer scope for diversified agriculture. Trade liberalization provides opportunities to reach for outside markets, which were not accessible otherwise. Great gains are possible through agro-based industrialization and overall rural development.

The cultivators and farmers of India have a rich heritage of traditional agricultural wisdom on crop and livestock husbandry and fisheries; based on the principles of the conservation of natural resources and environment. India has an extensive network of extension channels for dissemination of technologies generated by research institutes. India has a strong manufacturing capability and a large network of about 20,000 manufacturers and nearly one million village artisans for indigeneous production of all types of agricultural machinery. Therefore, inspite of several disadvantages, we have a great chance to leap forward, if we enforce the right policy initiatives and implement them.

Biotechnology and India Agriculture

We need to recongnise that, just as the food reequirements of today's population of nearly 6 billion people could not have been met by the technologies of the 1940's, we cannot assume that current practices will feed the population of 8 billion expected by 2020. New approaches are needed in addition to the continued improvement of existing methods of crop and animal husbandry and food processing.

The strategic integration of biotechnology tools into India agricultural systems can revolutionise Indian farming and usher in a new era. Genetic engineering is clearly the most revolutionary tool to impact agricultural research since the discovery of genetics by Mendel. Prior to genetic engineering, the exchange of DNA material was possible only between individual organisms of the same species. With the advent of genetic engineering in 1972, scientists have been able to identify specific genes associated with desirable traits in one organism. For example, a gene from bacteria, virus or animals may be transferred into plants to produce genetically modified plants having changed characteristics. This method therefore allows mixing of the genetic material from species that cannot otherwise breed naturally. Genetic engineering can be vital for an agrarian country like India. It can help in minimising the crop damage through disease and pest resistant varieties, reducing the use of chemicals, enhancing stress tolerance in crop plants, thus permitting productive farming on unproductive lands, etc.

One could even extend the growing season of crops and minimise losses due to environmental factors on the one hand and increase the shelf life of fruits and vegetables, on the other, thus minimising losses due to food spoilage.

We can expand the market vista and improve food quality. Biotechnology can also produce plants that possess healthy fats and oils, possess increased nutritive value, and create a whole range of higher value feeds. Biotechnology has even the power of producing biodegradable plastics, edible vaccines, etc. It is only through the blending of the 'gene revolution' with our experience in the 'geen revolution' that we can reach our goal of 'evergreen revolution' and also 'nutritional revolution' The advantage of the gene revolution is that it is relatively scale neutral, benefiting big and small farmers alike, It is also environment friendly.

Thus, it can be of great help to the smallest farmer with limited resources in increasing farm productivity through the availability of improved but powerful seed. It can also reduce a farmer's dependency on chemical inputs such as pesticides and fertilisers. Modern biotechnology offers unlimited opportunity for enhancing genetic potential of crops and other commodities, management of biotic and abiotic stresses, bio-remediation and organic recycling. India is one of the main centres of agricultural biodiversity and its gene richness can greatly complement the developments in modern biotechnology. Likewise, new developments in GIS, remote sensing, and crop modelling, provide new opportunities for integrated mana-gement of natural resources.

The revolution in informatics provides opportunities for sharing latest information for research and planning in a highly organised and efficient manner. With the globalisation of economy, the opportunities for value addition and post harvest management are also immense. With clever blending of technologies, the India farmer can usher in new era of confidence and performance.

Sir Francis Bacon had once said:

> *'It would be an unsound fancy to expect that things which have never yet been done, can be done except by methods which have never been tried'*

In the new context, we have to review the new role of biotechnology. Also, it is in this context that we have to view the rapid advances that modern biotechnology is making, including the area of transgenics. Let us focus on some of the developments in this area, since considerable interest and controversy has been created around the world today in this exciting area.

Movement in Transgenics

The advent of genetic modification over the last two decades has enabled plant breeders to develop new varieties of crops at a faster rate than was possible using traditional methods, with huge potential for further beneficial development.

Tobacco was the first plant to be genetically transformed\, in 1983, with cereals beginning in 1990, but it is only recently that products such as GM soya have reached the market place in Europe, and, for a variety of reasons, given rise to concern & controversy.

The concern must be addressed seriously if society is to exploit the new technologies appropriately. We will address this issue rather extensively a little later.

Whereas in India, we have been discussing and debating the issues, the rest of the world is galloping ahead.

Table: *Ag biotech product sales*

$ Millions	***Sales***			***Annual Growth Rate***	
	1994	1997	2002	1994-97	1997-02
Transgenic seeds and plants	20	405	2,030	173%	38%
Animal growth horomes	80	225	405	31	12
Biopesticides	35	65	110	23	11
Other	107	180	340	19	14
Total U.S. Sales	$242	$875	$2,885	53%	27%

Development, field testing, and commercialisation of transgenic crops, is now recongnized by the international scientific community to be an essential strategy for food security. Fortyfive countries, including India, have conducted transgenic crop field trials up to 1997.

The activity in the developing countries has been shown in Table.

Table: *Releases of transgenic plants in developing countries (by species and introduced trait)*

Species	***Introduced Trait***					
	Total Fied	***Herbicide Resistance***	***Insect Resistance***	***Virus Resistance***	***Product quality***	***Others releases***
Maize	46	29	16	1	3	5
Soya bean	27	25	—	1	1	—
Cotton	24	16	15	—	—	—
Tomato	19	—	2	1	16	—
Potato	13	—	1	6	—	6
Subtotal	129	70	34	9	20	11
Other species	30					
Total						159

The emphasis has been mainly on traits for herbicide, insect, virus & fungal resistance, and on product quality with a limited number of genes. Out of a total 159 releases. Argentina leads the way with 43, followed by chile, Mexico, Puerto Rico and the Republic of South Africa, with 17-20 each. India is far behind with just a few releases. There are fears relating to transgenic plants in society. These need to be addressed, not by brushing them aside, but by careful scientific analysis and educational programmes. All biosafety protocols must be followed rigidly. Let us deal with this issue more exhaustively now.

Public Concerns

The potential of biotechnology as a method to enhance agriculture productivity in the future has been accepted globally. However, because of its revolutionary nature, risk and uncertainty may be created by the process of genetic engineering and by the resulting genetically modified products. This may result from the introduction of a new unrelated DNA sequence into a recipient organism. The introduced DNA may have some unexpected effects on the cellular processes of the recipient organism. In addition to this, introduction of antibiotic resistance genes, as selection markers, pose serious implications in public health especially in genetically modified plants that are directly used for food purposes.

Risks are also associated with the genetically modified plants that are released into the environment. The nature of interactions with other organisms of the natural ecosystems cannot be anticipated without proper scientific testing. For example, modified plants with enhanced resistance to pests or disease threaten to transfer resistance to the wild relatives, which will have implications for biodiversity and ecosystem integrity. These, and many more, doubts plague the minds of common people.

Therefore, it is very clear that if the biotechnology potential has to be completely realized, it has to be done in an extremely responsible manner. At each step, proper testing and scientific data have to provided. Not only this, proper regulatory and policy mechanisms have to be put into place. It is therefore, absolutely essential, that conscious efforts be made especially by the developing countries to initiate well-defined programmes for the development and regulations of genetically modified plants.

To make this possible and to benefically, ethically and sustainably reap the benefits of biotechnology, there is a pressing need for scientists, researchers, policy makers, NGOS, progressive farmers, industrialists, and representatives of the government, to come together on a common forum in order to discuss the common concerns and find solutions to them.

Will modified plants transfer their introduced genes into wild relatives growing nearby?

Will modified plants that produce new compounds disrupt the 'balance of nature' in some way?

Could the planting of a restricted number of cultivators leave crop plants more susceptible to disease?

Could the planting of a restricted number of cultivators lead to a reduction in biodiversity (of crop plants, weed species, insects and microflora in the fields in question)?

Will genetically modified plants be able to avoid the factors that regulate natural populations and thereby change the usual ' balance' between populations?

Similarly, there are concerns about the social and economic effects. For example, some of these concerns are:

- How will the structure of farming (particularly in developing countries) be affected by biotechnology?
- How will patent laws affect traditional breeders 'rights (e.g. the right to save seed from one year to the next)?
- Will plant breeding be left increasingly in hands of a few companies, and if so, what effects might this have?

- Will some countries be 'plundered' for their genetic resources?
- We also have a set of ethical and moral issues that need addressing. In particular,
- The consumer has the right. What will consumers be told about the new food products?
- Is it acceptable to 'interfere' with nature through genetic engineering?
- Do we have the right not to use all means available to improve crop plants, especially when so many people are under or malnourished?

Then there are several regulatory issues. Among these, the three key issues are:

- Do current regulations give sufficient protection to farmers, consumers, those who have invested in, and those engaged in research?
- Is there sufficient international legislation to ensure environmental protection?
- Do current regulations compromise the competitiveness of biotechnology companies by being excessively restrictive?

The Royal society of London had appointed a group of experts to examine various aspects including the scientific evidence concerning the risk of transfer of genes from Genetically Modified (GM) crop plants to wild species and non-GM crops, the uptake of genes from GM food by the digestive system, and the current state of the regulatory system.

The experts concluded that the chances of gene transfer happening are slight, provided the regulatory processes are followed, but that this must be kept under consideration. The experts also concluded that the uptake of genes via the food chain is not a new issue because genes (i.e. DNA) are normal constituents of the human diet. Many products from GM Plants, such as sugar prepared from GM tomato paste, are so similar that they are regarded as 'substantially equivalent' others, for example flour from GM soya, may contain a new gene or its product, although many of the purification processes involved in food production will destroy and DNA present in the raw material.

Some GM foods have been produced using an antibiotic resistance 'market' gene, which is a laboratory device designed to identify genetically transformed plants. The Royal Society has shown concern about the use of such genes in food products and it suggested that any further increase in the use of such markers in the human or animal food chain would be undersirable.

Gm crop plants have been produced to improve insect tolerance and virus resistance, and to include herbicide tolerance, so that transferred to non-target species of plants, and that the development of resistance by target pests is minimised, the regulatory authorities must be assured that any negative effects would be no greater than those resulting from conventional procedures; and that any long term effects on the environment and ecology would be closely monitored, with statutory restrictions in place to control marketing; and also that best practice advice was adopted by growers.

The reliance on a case by case approach to the legislation, may result in lack of analysis of the overall impact of the technology on agriculture and the environment, and of the long term effects of gm. Although mechanisms are already in place to regulate many individual aspects of GM technology, there is no means for looking at GM technology as a whole.

As independent, overarching, regulatory body is needed to span departmental responsibilities, monitor the enforcement of existing or future regulations, and strengthen the guidelines to growers of GM crops, such as those specifying the isolation distances between Gm and non-GM crops. The body should also review and monitor the membership of advisory committees and regulatory bodies.

The Indian Effort in R & D

Indian has set up a large network of research institutions and agricultural universities located in different regions and states of the county.

Biotechnological Applications Relevance to The Indian Farmer

Increased food production must keep pace with the country's population increase. Not only the quantity of food made available should improve.

To keep pace with the present population growth and consumption pattern, India's food requirement has been estimated to cross 225 million tons by 2005 AD. This would mean an annual agricultural growth rate of 6.7 percent, a daunting task considering the rapidly shrinking resource-base and fast declining input-use efficiency in major cropping systems. Not withstanding the impressive gain in the agricultural production, the vast agricultural potential still remains highly under-utilized. There are serious gaps both in farm yield realization and technology transfer, as the national average yield of most crop varieties of most crop varieties is low.

Not many superior varieties is mot pulses, many oilseeds and vegetable crops have been evolved in recent past. In case of vegetables the varieties which were imported or developed many decades ago are still being grown by the farmers. Their yields are low and their inherent ability to respond to improve production technology is also limited. Farmers have accepted bybrids in my vegetable crops such as cabbage, tomato and capsicum.

Research and Technology Transfer

The Central and State governments continued to invest heavilyu in research, hence the Indian public sector research system has increased in size in real terms but by mid 90s levelled off. The number of hybrids and varieties developed and released by ICAR Crop Improvement Projects and State Agricultural Universities in major and minor crops are enormous and have been well documented.

Private research to develop superior hybrids was initiated by few seed companies in mid 60s. By 1987-88, the number increased to over 12 and in 1997-98 it is estimated at 34. With the governments recognition of provate sector's performance capabilities and with "New Seed Policy" announced in 1988, the number of private companies involved in crop improvement programs have been steadily increasing.

Number of Private Companies Engaged in Crop improvement Program (1987-1998)

Table: *Number of Companies with Crop Improvement Programs*

Crop	*1987-88*	*1993-94*	*1997-98*
Pearl Millet	12	23	25
Sorghum	10	12	15
maize	06	15	20
Sunflower	10	30	20
Cotton	09	21	34
Hibrid Rice	-	4	6

For want of financial, infrastructural and trained human resources, most private companies specialize in breeding superior hybrids in few crops characterized by low volume, high value and amenable to hybrid technology. In addition to the ICAR and Agriculture University research system, about 13 private sector seed companies are actively involved in vegetable R&D. Some of them have joint ventures with multinational or companies specializing in vegetable crop research and utilize research facilities available with their overseas partners. Few have created facilities for field testing imported and public bred varieties. The role of private sector in vegetable research is primarily in the development of F hybrids and few open pollinated varieties.

Table: *Hybrid Vegetable and Melon seeds of Private R&D available in the market (1996-97)*

Crop	*No. of Companies*	*No. of hybrid Varieties*
Tomatoes	11	45
Brinjal	10	38
Chilli	8	27
Bhendi	13	25
Cabbage	11	40
Cauliflower	8	21
Ridgegourd	6	8
Watermelon	5	13
Muskmelon	2	7

Number of private/public hybrids in major crops in market during 1995 are compared with the number in 1997 in table.

Table: *Number of private and public bred hybrids in major crops in the market during 1995 and 1997.*

Number of Hybrids in Market

	1995		*1997*		
Crop	*Private*	*Public*	*Private*	*Public*	*No of Private Hybrids grown on 2% plus area*
Pearl Millet	50	04	60	06	14
Sorghum	22	04	41	05	6
Maize	57	03	67	03	12
Sunflower	47	05	35	06	10
Cotton	73	15	150	15	19
Hybrid Rice	4	4	8	4	-

In pearl millet, sorghum, maize and cotton number of hybrids introduced by private, companies has been steadily increasing. Many hybrids especially in paddy and cotton have regional adaptability. The number of hybrids which are accepted by the farmers and may be grown on 2 percent or more area is much lesser, so also the number evaluated in coordinated project trails or identified for release. Many hybrids recently supplied by the private sector embodies new technology which is either imported or developed by private companies. Private R&D's real investment in research has quadrupled between 1986 and 1997. Subsidiaries or joint ventures with multinational companies account for 30 percent of all private seed industry research in 1997. Both domestic companies and multinationals companies have started research programs on rice and mustard. Recently, Monsanto has also focused on hybrid wheat.

Few private sector companies have also initiated joint ventures (e.g. Mahyco-Monsanto Biotech, ProAgro-Agro-Evo) to transfer specific genes to their product lines to provide effective protection against pests. Some of them have taken initiative to treat their seed with hightech chemicals to provide protection against sucking pests (aphids, jassids, thrips) during seedling and grand growth crop stages and diseases (downy mildew in bajra) during flowering.

Commercial Seed Production

Majority of certified/commercial seed is produced under contract with trained seed multipliers in suitable agro-climatic conditions. Both public and private companies make contact with progressive farmers or trained seed multipliers and negotiate terms and conditions of production.

Individual seed multipliers have small areas to contract and this greatly increases the cost of production and administration. To overcome this situation, contracts are arranged through local organizers, who in turn contract individual farmers, consolidate individual fields intoblocks and supply to the company the ultimate product.

Companies generallyprepare a technical package for seed multiplication specifying the kind of soil needed, isolation distances required, staggered plantion in case of a hybrid if needed, distances among rows and plants, depth of planting, time of nutrient and spray applications, frequency and time of irrigation, mode of harvest especially in case of hybrid seed etc. They prefer to contract between 300-400 acres in a village and designate such an area as a "seed village". They create seed processing facilities and place technically trained staff to guide seed multipliers, monitor actual production and supervise seed processing. Generally one such unit serves a cluster of 5-10 seed villages

Only seed lots meeting genetic Purity, Physical Purity, germination and disease standards prescribed by the government of India, are packed and commercially sold. User-farmers have become very conscious not only of genetic purity and germination, but also of physical attributes of seeds. They prefer bold, shining and plump seed, unaffected by mould The companies therefore, organize seed production in areas with appropriate agro-climatic conditions and no rains in the post-flowering period of seed crop.

They also ensure proper cleaning to remove undersized, premature, broken seeds or inert matter on cleaner-cumgraders, gravity table, de-stoners and thus improve physical attributes of seed lots. Seeds are treated with recommended pesticides to prevent them against stored grain and soil pests. To add value to their seed, many companies now treat them with specific chemicals to provide protection agains sucking pests during seedling stages or specific diseases such as downys mildew for 40-45 days after emergence.

The seed production of most hybrid crops is generally undertaken in Western and Southern parts of the country because of congenial agro-climatic conditions and possibility of taking two seed crops in a year if necessary. The prominent states are Gujarat, Maharashtra, Andhra Pradesh, Karnataka and Tamil Nadu. In contrast, the Northern states from Punjab to West Bengal are dominated by rice, wheat, and legume crops which are less profitable. Consequently there is a major difference in the cadre of seed industry between the North and West/South of India.

Technology of seed production is moving at a rapid pace. Hence the companies have to ensure that they pass on the technology to the contract seed multipliers to harvest high yields of good quality seeds. Most public bred hybrids and variety seeds are certified while the proprietary hybrid seeds are sold under truthful label. All seeds meet minimum prescribed standards but for proprietary hybrids quality monitoring is through their in-house system.

Private Hybrid's Share of Seed Market

Private sector has played a major role in supplying quality seeds of both private and/or public hybrids. It is interesting to note that 69% of the pearl millet seeds produced during 1996-97 season was of private hybrids. Other crops where private research plays key role in seed supply are sudangrass sorghym (100%), sunflower (98%), Maize (90%), sorghum (52%) and cotton (50%). The contribution of private research in value terms is also steadily increasing. The share of research hybrids to total turnover in these crops was about 70 percent in 1996-97 as compared with 46 percent in 1990-91. Private hybrids are successfully meeting the regional needs of early maturity, tolerance to pests, unpredictable agro-climatic condition and soil variations.

Government Intervention: Legislation and Policy Issue

At the highest level, government policy is to reduce controls and to make the economic system more transparent in practice however, there is potential proliferation of legislation affecting seed sector. Given the current climate of economic development there is an urgent need for positive relationship between the government and the private sector.

In order to fulfil farmer's needs and GOI food production targets set for the year 2005, the existing misapprehensions should disappear to ensure private sector investment in infrastructural improvement and expansion to increase quality seed production.

Presently Indian seed industry is governed by the following Acts:

- Seed Act 1966
- Seed Rules 1968
- Seed Control Order 1983
- Essential Commodities Act 1955
- Package Commodities order 1975
- Standrads of Weights and Measures Act 1976
- Consumer Protection Act 1986.

Export Regulations and Quarantine :

- Plants, Fruits and Seeds (Regulation of import into India) Order 1989
- Urban Land Ceiling Act 1976
- State Acts of land acquisition/land use
- State Acts for the control/movement of crops and seeds.

The policy framework hitherto has been regulatory with excessive legislation. Above plethora of legislation covering the seed sector is out of date and restrictive rather than progressive. The multiplicity of Acts, rules and administrative orders have resulted in over regulation of the nascent industry.

Future Trends

During past four decades spectacular progress has been achieved in agricultural production in the country. Compared to 51 million tons of food grain production in 1950-51, India realized a food grain production of over 196 million tons during 1997-98, resulting not only in self-sufficiency in food but also in a buffer stock of over 30 million tons. Food grain production is expected to cross 200 million tons in current year (1998-99). In the oilseed sector, which is primarily production was coupled with important policy planning and research and technology development transformation.

The steady increase in productivity has been accomplished by exploiting hybrid vigour in maize, sorghum, pearl millet, sunflower, cotton, castor and improved genetic potential of high yielding varieties in wheat, paddy, many legumes and oilseed crops. The superiority was for high inherent harvest index, inbuilt resistance to major disease-pest complex, ability to withstand environmental stress conditions, adjusting growth cycle to better suit the available growing conditions, generally greater physiological efficiency and/or improved quality. Nineties focused on exploiting heterosis in rice, rapeseed-mustard and cajanas for enhanced productivity, sustainability and utilizing the vast agricultural potential which still remains highly under-realized through bridging the serious gaps in inherent yield potential and ensuring a rapid technology transfer to increase national average yields leading to increased produ-ctivity.

The Indian seed industry shows a clear pattern of development that also reflects in government policies. The decade, after the announcement of New Seed Policy of 1988, witnessed entry of major multinational companies and foreign genetic material to the Indian seed market. Nineties also saw proliferation of domestic seed companies into more dynamic prevailing economic climate. Although public sector dominance continues in open pollinated variety seeds, the private sector is steadily increasing its market share of hybrid seeds. Commercially the seed industry is going through a period of accelerated development with activities now including in-house research, adoption of agro-biotechnology and genetic engineering techniques.

Till such time that the population stabilizes, the demand for food would continue to increase. To keep pace with the present rate of population growth and consumption pattern, India's food requirement has been estimated to cross 225 million tons mark by 2005. The certified/quality seed distribution has considerably increased from a mere 18,300 tons in 1952-53 to over one million tons in 1997-98. The quality seed requirements are estimated to increase to 1.8 million tons by 2005 AD. If the population growth continues at current rate, the country could require 300 million tons of food to feed an estimated 1.4 billion people by the year 2025.

This will only be possible with an average agricultural growth rate of about 5 percent annually. India, therefore, need to continue the green revolution through introduction of value added superior genetic material possessing inherent high yield potential even under environmental stress conditions and genetically engineered or chemically induced seed treated tolerance to major disease-pest complex. Then only the income levels of small and marginal farmers who dominate Indian agriculture today will rise. Accomplishing such targets of production will not be an easy task considering the fast declining input use efficiency for major cropping systems and rapidly shrinking resources base.

Future agricultural development would be guided not only by the compulsions of improving food and nutritional security but also major concerns for environmental protection, sustainablity and profitability. Following GATT agreement and the liberalization process, globalization of market would call for competitiveness and efficiency of agricultural production. Projected population migration from rural to urban areas will also demand a significant shift in the pattern of the food basket. Developments in biotechnology, molecular biology including genetic engineering and new cutting edge technologies in recent years will provide new potential for accelerated growth.

Used in consort with conventional approach and harnessed judiciously, these new break throughs would help increase agricultural productivity and sustainablity. To meet the future challenges of increased productivity-a daunting national need, the problem oriented, demand driven, value added approach of integrated research and development involving all reasonable players as partners would offer practical solution.

This would necessitate an active cooperation between the public and private solution. This would necessitate an active cooperation between the public and private sector agricultural research, massive private invest-ments in R&D, global exchange of germplasm and technology and necessary changes in government policies to create a congenial environment to encourage entry and active participation of technology rich foreign private companies in Indian seed market. Investments and technology would needed to breed hybrids and varieties with high inherent genetic yield potential with value added traits, incorporated through genetic engineering to improve nutrition, provide protection against major pest-disease complex, inbuilt drought tolerance to ensure increased productivity even for arid and semiarid zones of the country. Biotechnology applications would not only minimize loss of biodiversity but will also expedite introduction of high yielding traits in desired crops of major economic importance.

ICAR is currently preparing a new perspective plan to upgrade farm research and man power development systems in the country. They are being reoriented to ensure sustainability and globalization of agriculture with emphasis on diversification, value addition and to encourage exports.

This is opportune time for bold initiatives and policy decisions in support of private R&D which is steadily becoming stronger. The number of proprietary hybrids is increasing rapidly. Hybrids in rice are being commercialized. Hybrids in rapeseed-mustard and cajanas are on threshold of commercialization. Research efforts to evolve hybrids in wheat are on.

Innovations in genetic engineering are being incorporated in Indian products; in cotton to provide resistance to Heliothis bollworm; in brinjal, cauliflower and cabbage among vegetables for black moth. Genetically engineered products will probably enter commercial market in early next decade. The present value of formal seed market estimated at Rs. 20,190 crore, i.e. US$ 480 million with private sector contributing approximately 72 percent, is likely to double in next 3-4 years. There is lot to be gained from these synergies that remain untapped between government laboratories, industrys and academia. It is an appropriate time for an effective partnership between ICAR projects and institutes, state agricultural universities and private R&D to evolve.

Time is ripe for a second green revolution which would not only concentrate on increased productivity but also on value added traits to reduce cultivation cost, pollution dangers and improve quality of products. Rapid developments in biotechnology, particularly in genetic competition. Indian seed industry is ready to face the challenge. Government is also cognizant of need to remove bureaucratic hurdles to its expansion. It is most opportune time for government to take bold decisions in favour of private sector development. It needs to start recognizing and rewarding innovations and encourage domestic companies to invest greater resources in research. Amended Indian Patents Act Of 1910 includes biotechnology and genetically engineered product patents and has been Passed by Rajya Sabha (Upper House). Government should expedite implementation of Plant Variety and Farmer's Rights Protection Bill. These steps would send right signals to the international of Plant Variety and Farmer's Rights Protection Bill.

These steps would send right signals to the international seed market and would encourage massive investments both from multinational and domestic seed companies to set up high-tech R&D, seed production and conditioning facilities to produce seed meeting international standards, for both domestic market and for export. Seed industry needs to become technology driven and innovation led. This transition will have to be facilitated through innovating strategies where both government and industry are partners. The private sector's R&D emphasis on hybrids would also then include open pollinated varieties in major self-pollinated crops where hybrid systems are not yet identified. As the business expands, large companies would create a full line of products to meet their dealer's demands. Since most of this technology will be high value proprietary products, many farmers are likely to become contract growers. Their risk will be limited, and profits although smaller will be guaranteed. New technology thus holds the potential to make farming more secure and profitable.

The guiding philosophy of ICAR research for next millennium's hi-tech environment should be to complement private sector R&D efforts. ICAR's "research product mix" should reflect priorities and targets that continuously evolve in response to changing internal and external research environment. Its policies should consider a shift in emphasis to upstream research producing mainly intermediate outputs, i.e. inputs to further research, which in turn will yield better products. In past 4-5 years over 19 billion US dollars have been committed to mergers, acquisitions and technology alliances in the crop genetic industry world wide.

There has also been incredible growth of big com-pany R&D budgets in this area. The agriculture in the developing world is becoming like any other industry and is being dominated by a hand full of big technology developers. Until a decade ago hundreds of competing companies except for one or two big ones accounted for nearly half of total seed market share. Now the technology is a hand full of large companies developing the technology and controlling much of the industry and fewer smaller players. This worldwide changes would have a major impact on the pattern of Indian seed industry growth in the coming century.

Seed industry worldwide is also going through a rapid technology revolution with value added output traits at its centre, driven by consumer and commercial end users. It involves all aspects of agribiotechnology and crop production. Emphasis is on proprietary genes in value added crops for the feed, food, chemical and pharmaceutical industries at the centre of new seed industry.

This change to value added agriculture will also have major impact on restructuring of Indian seed sector, hence seed business in coming decades. It will become more difficult for seed companies to assess new technology. In the changing seed industry scenario, seed companies in India will continue small-scale plant breeding, product testing and maintaining a strong, localized seed distribution network with cordial customer relationship.

Using this strength they can get involved in alliances and agreements that may ensure access to the future technology. The future trend is quite clear from Monsanto's equity acquisition in Mahyco; Mahyco-Monsanto Biotech Ltd.; Mons-anto acquiring Cargill and buying equity in E.I.D.Parry's seed business; Agr-Evo acquiring Pro-agro; Semenis buying 40 percent additional equity stock in Nath Sluis and so on. Hence the Indian seed companies will continue to play a basic and important role.

Like present day contract grower farmers, small seed companies of all sizes will incre-asing become contract producers and developers for the bigh technology rich companies. Because of rapidly changing new technology the seed industry world over is going through a radical change with a whole new future of rapid growth ahead of it. Indian seed sector would have no option but to follow that path with active government support and huge private investments wherever needed. A large cadre of highly skilled seed multipliers exist. Companies both in public and private sector have developed systems for training technical man power to monitor expanded infrastructure.

As their R&D gains strength, they will develop new hybrids and varieties suitable not only for Indian market but also for the agro-climatic conditions prevailing in the export market. India's low cost skills at all levels of seed industry from research through seed production coupled with range of environments suitable for production of most crop seeds and rapid flow of technology offers excellent base for competitive supplies in the international market. Indian seed industry has reached a stage of maturity and is in a position to shape itself to compete at national and international levels.

This may however, be recognized that private sector for some more time will continue to concentrate on more lucrative sector of the market, with a strong emphasis on hybrids and the more specialized low volume crops, e.g. vegetables. Fundamental strength and potential of Indian seed market cannot be doubted. The country has a very large population and a strong commitment to self-sufficiency in food production. This can only be achieved by increasing productivity per unit area.

Indian economy is opening up the opportunities and challenges offered by the globalization of agriculture, particularly in the seed sector. Given that the prevailing uncertainites about legislative changes which currently retard seed industry growth are favourably clarified, the present trend of industry's accelerated growth in domestic market, value addition to their products, custom production and export will continue. Indian seed industry is on the threshold of becoming a global player. Beginning next century, Indian private sector can look forward to playing a key role in the development of Indian seed industry and take its rightful place in the global market.

5

Agriculture Biotechnology in the Forestry Sector

"Biotechnology provides important tools for the sustainable development of agriculture, fisheries and forestry and can be of significant help in meeting the food needs of a growing and increasingly urbanized population," reads an FAO press release dated 15 March 2000. The field of modern biotechnology is indeed often considered as one of the fields of scientific research in which the most rapid advances have been made in recent years. Several elements can explain the growing interest of forest scientists, conservationists and tree growers in modern biotechnologies.

They include the unique roles and functions that trees, major structural constituents of forest ecosystems have, their special biological characteristics, and their importance in the provision of environmental, social and economic goods and services.

Special features of interest to scientists and geneticists include the low level of domestication of forest trees and their rich genetic diversity; their long life cycles, long generation times and late sexual maturity; their spatial requirements; the multiplicity of species and the low degree of heritability of traits of interest, linked to weak juvenile–adult correlations and the importance of genotype–environment interactions.

Application of biotechnologies in forests has been seen as a unique opportunity for obtaining new information on the extent, patterns and functioning of tree genetic diversity; and for providing new tree varieties and reproductive materials adapted to changing environmental, social and economic environments.

Background and Methodology

Specific developments in biotechnology in the forestry sector have been addressed in a large number of conferences, meetings, publications, electronic fora and Internet web pages. Owing to this abundant literature, this report, commissioned from CIRAD-Forêt by FAO in December 2003, will not describe the types and classifications of forest biotechnology in detail.

It aims instead to fill a gap in global data and statistics on research in and applications of biotechnology for forest trees. Given the scientific and technical potential created by an increasingly accurate knowledge of forest tree species genomic structure, it is important to have an overall picture of the current status of forest biotechnology developments, together with trends and future prospects. The objective of this document is thus to review and summarize research, and the suitability and practical use of biotechnology in the forestry sector, and to provide tentative global analyses.

For the purpose of the study, a simple data set has been developed. The data set gathers major biotechnology activity (i.e. a given technology developed or used in a given country, on a given tree species or variety, by a given laboratory team, for a given purpose). Data originated from (i) systematic searches in CAB Abstracts and associated global scientific databases, (ii) searches on the Internet (including sites of private companies, governmental and non-governmental institutions and linked references, and (iii) personal enquiries, observations and communications. Most significant publications, including those produced by major laboratories and teams, have been included in the data set. The study was mainly conducted between February and September 2004.

The data set included basic fields such as country, type of biotechnology, information source, reference or Internet site, stage of development, species or genera involved and, whenever available, year when the activity was conducted. These fields were considered to be a minimum set of requirements for entering a biotechnology activity in the database. The reference period covers approximately the last 10 years, although more than 75 percent of the data were from between 2000 and 2004. Internet references, however, could not always be dated.

The data set is in no way comprehensive, and some of its limitations reflect the difficulties of such information gathering. The study revealed language limitations (international databases cover only a fraction of the literature in Chinese and Russian, for example). International databases also tend to reflect past research activities, and only a small share of on-going research work. An additional flaw of the data set is related to private (corporate) research and commercial applications, for which public-domain information is generally scarce. Despite its drawbacks, the data set provides a sample (of unknown global representativeness) of materials available in the public domain. No attempt has been made to gather classified information. In total, data on 2 716 activities were collected, and their analysis supports the present report. Data, statistics and conclusions presented in this report should therefore be considered with caution and as general indicators.

Outline of the Report

A quick review of the literature shows that various classification systems have been used to categorize modern biotechnologies used in forestry. In some countries, the term biotechnology refers to genetic modification exclusively. Yanchuk (2001) proposed three main categories: (i) biotechnology employing molecular markers, (ii) biotechnology aimed at enhancing plant propagation and large-scale production of uniform plant material, and (iii) biotechnology for modifying the genome of forest tree species. In the present report, the broad categories include:

1. The characterization of genetic diversity of forestry species, including diversity structure studies, gene flow and human impacts on forest stands: characterization, population genetics and diversity studies.
2. The functional and applied aspects of genetic investigation: genetic mapping, marker-assisted selection (MAS) and genomics (collectively referred to as MMG).
3. Research and applications in vegetative propagation: micropropagation (in the broad sense).
4. The alteration of genomes by insertion of genes: genetic modification. Individual data (biotechnology entries) were categorized according to the above classification once they had been entered into the data set.

The report presents detailed analyses of the data sets for the categories 1–4 above, and provides separate summaries for biotechnologies excluding genetic modification, and genetic modification technologies.

Summary of the State of Biotechnology in the Forestry Sector, Excluding Genetic Modification

The information compiled in the data set of biotechnology activities excluding genetic modification represents 2 196 references (or 81 percent of all activities reported). Activities were reported in 76 countries, broken down by regions as follows: 39 percent of the activities were reported in Europe, 24 percent in Asia, 23 percent in North America, 6 percent in Oceania, 5 percent in South America, 3 percent in Africa and less than one percent in the Near East. Both developed countries (24 countries, representing 68 percent of biotechnology activities) and developing countries and countries in transition (52 countries, or 32 percent of activities) were represented.

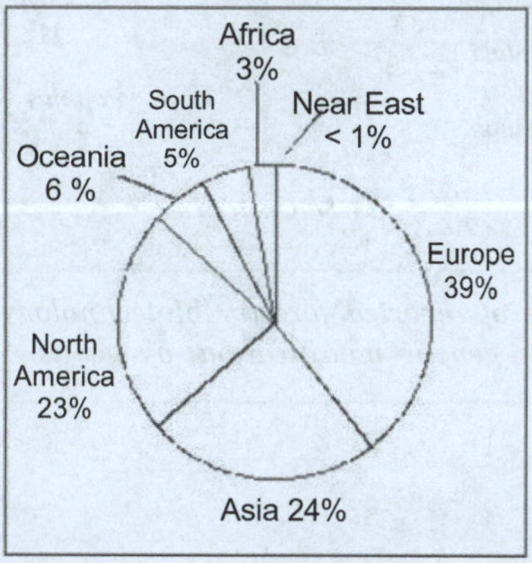

Figure: *Distribution of reported forest biotechnology activities (excluding genetic modification) by world region.*

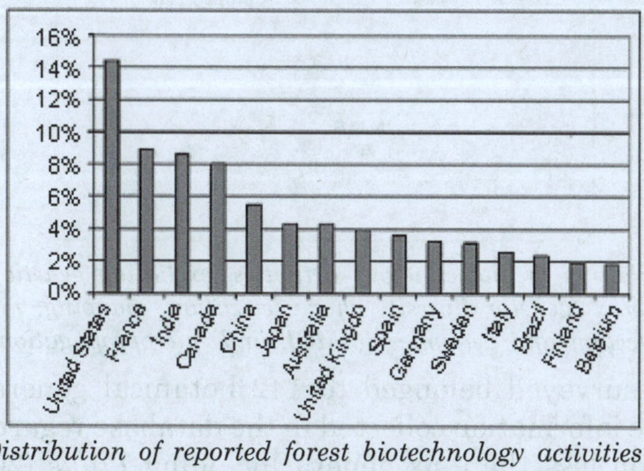

Figure: *Distribution of reported forest biotechnology activities (percent of activities in the data set, excluding genetic modification) by country (for the 15 countries most represented, making up 77 percent of the data set of entries excluding genetic modification).*

Developing countries and countries in transition were mainly represented by India (27 percent of these countries' activities), China (17 percent), Brazil (7 percent), South Africa (5 percent) and Malaysia (4 percent). Three countries (India, China and Brazil) accounted for 52 percent of all biotechnology work reported in developing countries and countries in transition.

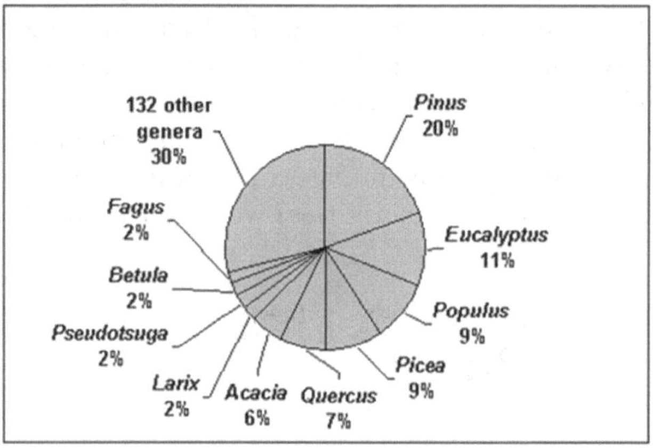

Figure: *Distribution of reported forestry biotechnology activities, excluding genetic modification, by genus.*

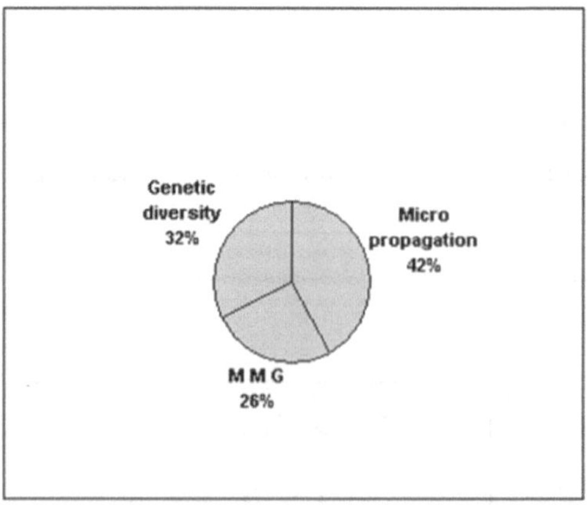

Figure: *Distribution of biotechnology activities, excluding genetic modification, by broad category (genetic diversity characterization; mapping, marker-assisted selection and genomics [MMG]; and micropropagation).*

Species surveyed belonged to 142 botanical genera. Sixty-two percent of the information collected in the database regarded research carried out on less than six genera including *Pinus* (20 percent of biotechnology activities excluding genetic modification), *Eucalyptus* (11 percent), *Picea* (9 percent), *Populus* (9 percent), *Quercus* (7 percent) and *Acacia* (6 percent). Just four genera (*Pinus, Eucalyptus, Picea* and *Populus*) account for almost half of the compiled biotechnology activities excluding genetic modification. Work was found to be relatively evenly spread between the three main categories of biotechnology categories apart from genetic modification: characterization of tree species genetic diversity represented 32 percent of biotechnology activities, MMG 26 percent, and micropropagation 42 percent. Differences were more marked when tree genera were considered (Figure 3B).

The forestry sector appears to have rapidly adopted markers developed for agricultural crops (Figure 4). Isozymes and random amplified polymorphic DNAs (RAPDs) have been widely used for genetic diversity description although the present trend seems to favour microsatellites (nuclear and chloroplast) and amplified fragment length polymorphisms (AFLPs). Driven by research on genomics, expressed genome banks are being widely developed.

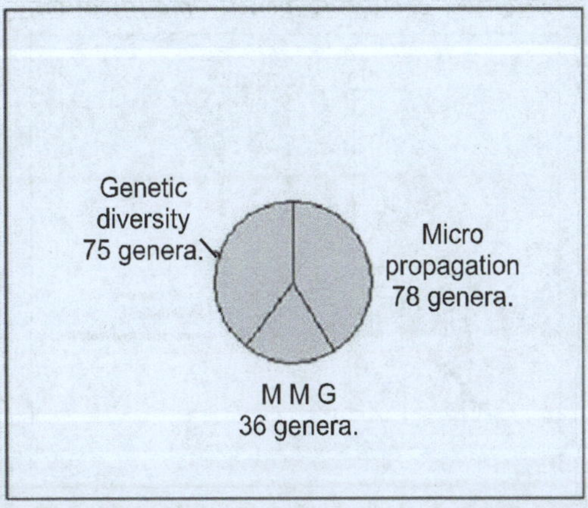

Figure: *Distribution of biotechnology activities, excluding genetic modification, (genetic diversity characterization; mapping, marker-assisted selection and genomics [MMG]; and micropropagation) by number of tree genera.*

The majority of the work reported is still mainly at the experimental stage in the laboratory. Genetic diversity characterization has less than one percent of its reported activities in the field, MMG 2.5 percent and micropropagation 5 percent (Figure 5).

Field tests are still mainly geared to supporting laboratory research. These results possibly reflect the origin of the information in the data set. While research activities in the public sector are relatively easy to collect, especially through international research storage databases, information on commercial applications is generally restricted and incomplete.

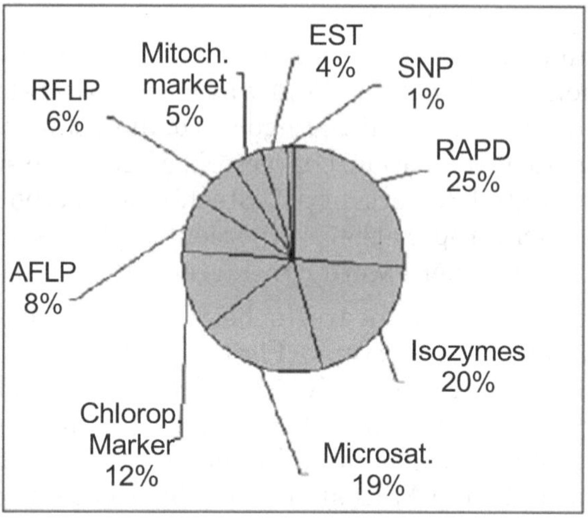

***Figure:** Distribution of molecular markers used in forest biotechnology activities, excluding genetic modification.*

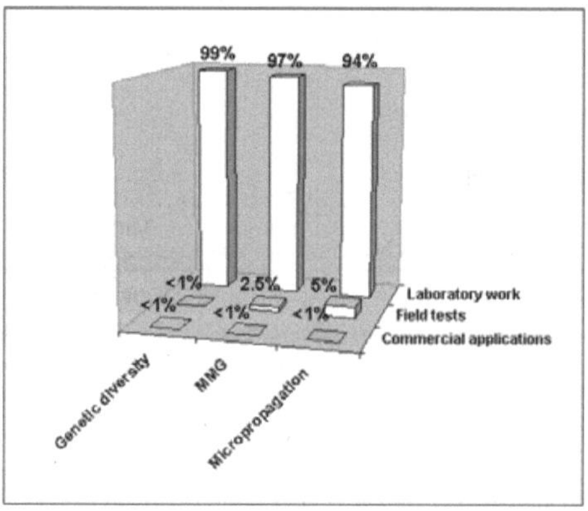

***Figure:** Distribution of reported forest biotechnology activities, excluding genetic modification, by category and applications (laboratory studies, field trials and commercial deployments)[1]*

*[1]**MMG:** mapping, marker-assisted selection and genomics.*

Commercial applications of micropropagation are, however, generating increasing interest. The potential is huge although, up to now, only several thousand hectares seem to have been established globally using micropropagated materials.

In South America, particularly in Brazil, some companies are reported to be integrating micropropagation into the clonal propagation process: micropropagation is used to 'store' clones in mother blocks (gene banks) in the laboratory as potential sources of responsive nursery stock plants for large-scale mass propagation. The use of rooted cuttings has allowed the propagation costs to be lowered significantly. Great expectations have been raised about the possible contribution of biotechnology to tree selection and breeding, and its commercial applications. Genomics and proteomics should greatly help breeders in tree selection, in particular in the identification of traits of interests. However, it remains difficult to predict when new forest tree varieties selected with biotechnology tools will become available on the market. Although genetic diversity characterization started some 30 years ago, very limited large-scale commercial application has yet been reported in forest tree genetic resources conservation and management.

Characterization of Genetic Diversity: Population Genetics and Diversity Studies

Genetic diversity is sometimes considered to be the invisible dimension of biological diversity. The present structure of genetic diversity is the result of the evolutionary history of the species exposed to natural selection pressures in variable environmental conditions. Natural selection at the local level is an evolutionary force opposed to gene flow.

The combination of the two forces creates a powerful mechanism for maintaining within-species diversity. The use of molecular descriptors (markers) of the genome has allowed the measurement of genetic variation between genotypes and within/between populations, as well as the effectiveness of seed and pollen dispersal.

Global Review

Studies carried out on forest tree genetic structure and on reproductive regimes use RAPD and AFLP markers, as well as nuclear markers and microsatellites, to increase understanding of the relationship between genome variations and genetic diversity. Molecular markers, especially neutral markers, provide an important tool for studying the structure of populations.

However, experience shows that the resulting information on diversity, for example in ecology or taxonomic studies, is not always a good indicator of the patterns of variation in traits subject to selection pressures. A complementary approach is often required, consisting of studying the genetic diversity in specific regions of the genome. For example, SNPs (single nucleotide polymorphisms) are used to study the impact of evolutionary forces on the variability of allelic forms of a given gene, and their phenotypic expression. Work reported on tropical trees often involves studies of the impacts of tree harvesting and fragmentation on species' reproductive patterns and changes in genetic diversity at the population level. In temperate species, several studies focused on gene flow in natural populations (*Quercus*, *Cedrus*) and the evolution of genetic diversity under forest management. Genetic diversity descriptions are numerous and diversified in terms of species and methodologies, and are mainly aimed at verifying some theoretical aspects, or at providing elements for the elaboration of conservation strategies.

Applications of results at operational level have been infrequently reported as yet, except in the case of rosewood (*Aniba rosaeodora*) and sandalwood (*Santalum*) (J. M. Bouvet, personal communication). On-going studies on gene flow and, more specifically, on pollen flow, are based on the use of molecular markers, mainly microsatellites and on paternity analysis. Objectives include:

- estimating and studying the evolution of genetic variability over time;
- assessing effective population sizes;
- studying biological mechanisms of reproduction, either in natural populations or in improved populations (in selection programmes);
- studying pollen pollution in tree seed production areas (seed stands or seed orchards) or in the context of GM trees.

Data Set Analysis

Most of the activities compiled in the data set have the overall objective of conserving and managing forest genetic resources in a sustainable way, and providing a better understanding of genetic diversity and evolution through the study of gene flow. This includes analyses of the consequences of human practices on the evolution of forest tree genetic diversity, as well as development and validation of theoretical models of population genetics.

Work has been reported from all regions of the world (Figure 6A) including Europe (44 percent), followed by North America (22 percent), Asia (19 percent), South America (6 percent), Oceania (6 percent) Africa (3 percent) and the Near East (less than one percent). When one considers the origin of the species being studied, the situation shows broadly similar patterns (Figure 6B). At least 99 genera are being worked on: *Pinus* (19 percent), *Quercus* (12 percent), *Eucalyptus* (8 percent), *Acacia* (7 percent) and *Picea* and *Populus* (6 percent) are the most frequently cited genera (Figure 7).

Reported objectives of genetic diversity characterization studies (Figure 8) include: evaluation of genetic diversity (57 percent), gene flow studies (15 percent), fingerprinting (12 percent), and conservation (6 percent).

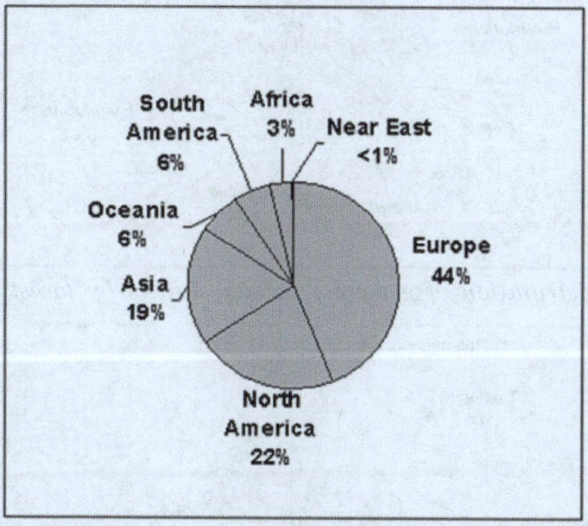

Figure: *Distribution of forest genetic diversity characterization activities by region where the work is/was carried out.*

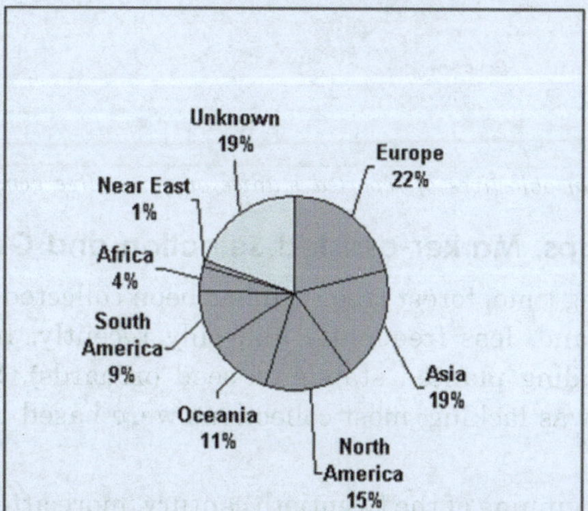

Figure: *Distribution of forest genetic diversity characterization activities by region of origin of the species studied; 'unknown' relates to undocumented sources, or to hybrid species, varieties and clones.*

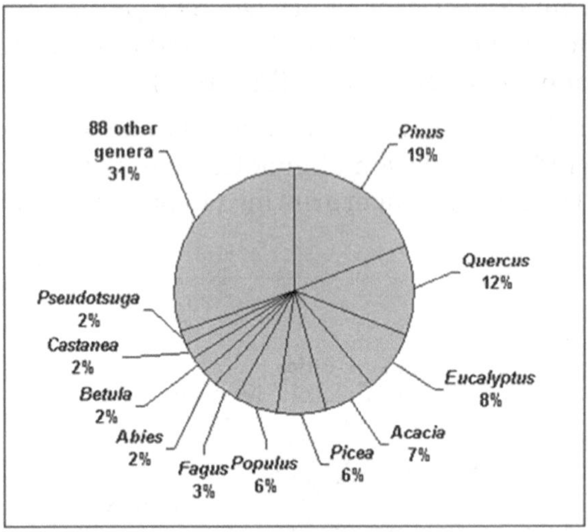

Figure: *Distribution of genetic diversity studies by forest tree genus.*

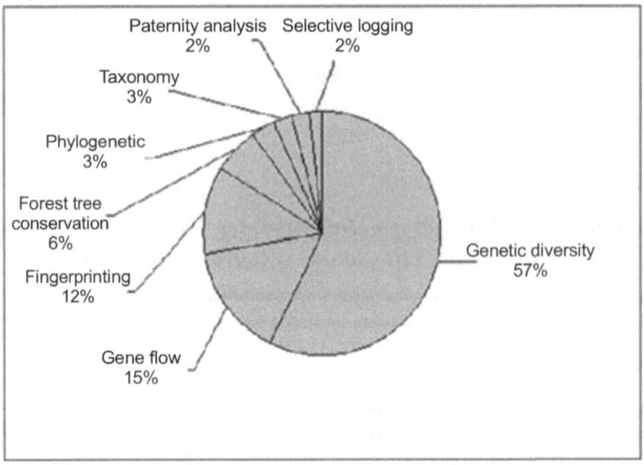

Figure: *Main objectives of reported studies on forest tree genetic diversity.*

Genetic Maps, Marker-assisted Selection and Genomics

For a long time, forest tree seed has been collected from natural populations and, less frequently and only recently, from artificial stands (including planted stands or seed orchards). Since genetic information was lacking, most collections were based on phenotypic selection.

At the beginning of the twentieth century, more attention started to be given to seed origin; numerous arboreta were established around the world and international seed exchange, documented to a greater or lesser extent, increased significantly.

Later, growing awareness of the importance of intraspecific variation led to a more systematic exploration of the natural distribution of species of proven or perceived interest. With the assistance of national and international organizations, comprehensive explorations were made in the 1960s and 1970s for several genera and species including *Tectona grandis, Gmelina arborea, Eucalyptus, Pinus* and *Acacia*.

Seeds collected were used in provenance tests, which in turn provided basic information and new materials for seed production areas or tree seed orchards. Nowadays, biotechnology tools allow more in-depth exploration of within-species genetic diversity, through the study of genomic variability.

Use of Ciotechnology in Tree Breeding and Selection

Marker-assisted selection (MAS) has given further impetus to tree breeding and selection. Molecular markers are genetically linked to a given allele on a given locus and can therefore be used to predict the presence of the allele with great accuracy. The first markers developed (RAPDs, RFLPs [restriction fragment length polymorphisms] AFLPs and microsatellites) were neutral in the sense that their function, if they had any, was unknown.

They have provided a way of estimating genetic diversity and allowed comparison between individuals in different growth and development conditions. They have also been used in profiling to tag and then single out individual trees.

Co-segregation of RAPD and AFLP type markers has allowed the construction of genetic maps. Genetic maps can be used to assess the degree of evolutionary relatedness of a number of species and identify the zones where gene variations are statistically linked to the variability of quantitative traits. These zones are called QTLs (quantitative trait loci).

QTL research focuses on the genetic architecture of traits of interest. However, QTL research results are not easily transferred to other tree populations: the QTLs identified so far are closely linked to the populations sampled. The use of QTLs in tree selection is thus increasingly questioned. For the above reasons, the use of candidate genes became necessary for applying MAS to forest trees. The candidate gene approach was made possible by the development of gene search methods (AFLP, EST banks and SSR, cDNA), transcriptomics and proteomics. In order to use those genes in selection, their molecular variability must be linked to the variation of the targeted trait of interest. This requires associating appropriate field experiments with genetic mapping.

Proteomic analysis offers an important field of basic research that many research teams are willing to develop. Proteomic studies can find applications in such fields as biotic and abiotic stress response, effects of genetic mutations, gene expression regulation, QTL validation and genetic variability.

Information technology is an important asset since work on biotechnology requires the handling of massive catalogues of genes, transcripts and proteins. The use of bioinformatics, in particular, to transfer results from one given (model) species to other species, in variability studies, fingerprinting or tree breeding, is particularly valuable.

Database Analysis

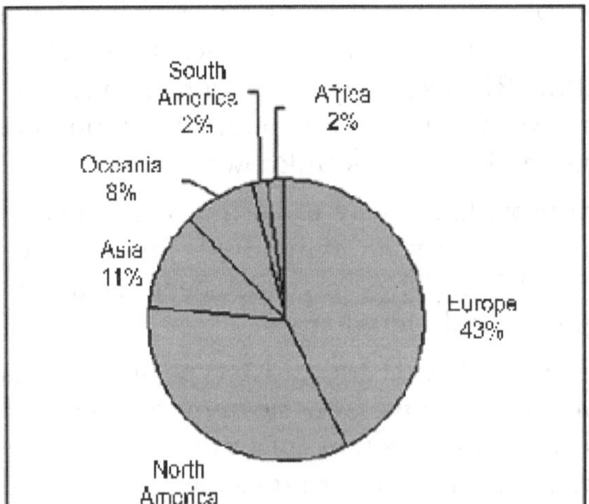

Figure: *Distribution of research related to mapping, marker-assisted selection and genomics (MMG) in forest tree species by region where the work is being carried out.*

Mapping, MAS and genomics (MMG) studies are found in all regions of the world except the Near East (Figure 9). Europe and North America represent 43 percent and 34 percent, respectively, of the total MMG activities, followed by Asia (11 percent), Oceania (8 percent), South America (2 percent) and Africa (2 percent). The work documented in Africa (South Africa and Congo) and South America (Brazil) relates to *Eucalyptus* almost exclusively.

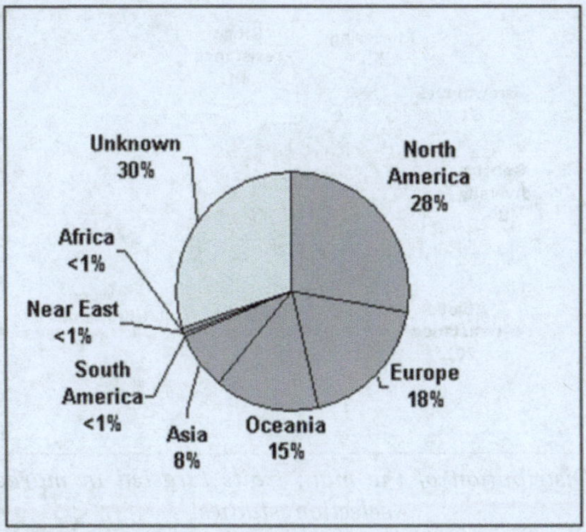

Figure: *Distribution of research related to mapping, marker-assisted selection and genomics (MMG) in forest tree species by region of origin of the species studied.*

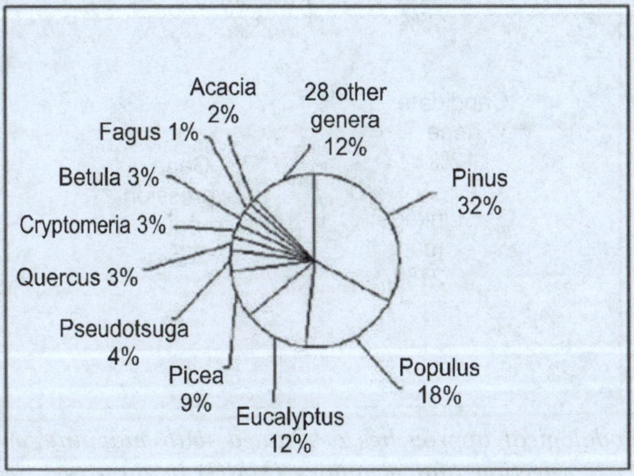

Figure: *Distribution of mapping, marker-assisted selection and genomics (MMG) activities, by genus..*

MMG activities are being currently applied to approximately 40 genera (Figure 10) among which *Pinus* (32 percent), *Populus* (18 percent), *Eucalyptus* (12 percent) and *Picea* (9 percent) are the most studied.

Objectives of the research work (Figure 11) include wood properties improvement in relation to lignin composition (57 percent), abiotic resistance for cold or drought stress (20 percent), genetic diversity (8 percent), growth rate (6 percent), flowering (5 percent) and phytosanitary aspects of biotic resistance (4 percent).

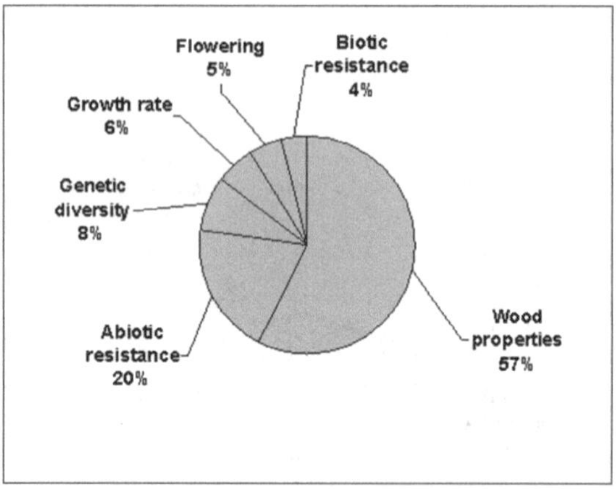

Figure: *Distribution of the main traits targeted in marker-assisted selection studies.*

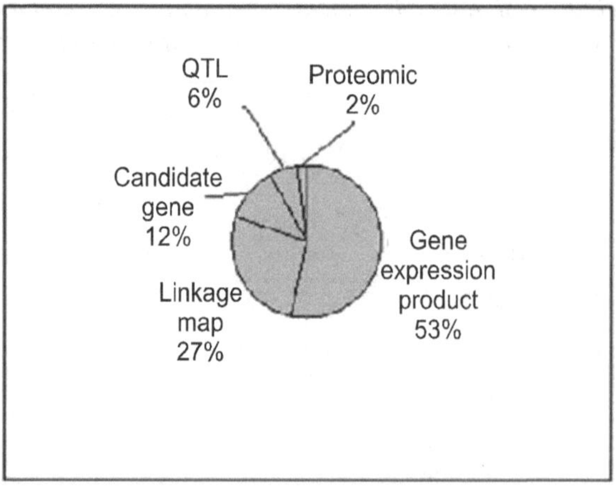

Figure: *Methodological approaches associated with mapping, marker-assisted selection and genomics (MMG) in forestry.*

The work is being pursued mainly on expressed gene products (53 percent), genetic map construction (27 percent), gene candidate research (12 percent) and QTLs (6 percent). More recent proteomics research constitutes 2 percent of the activities documented (Figure 12).

Genetic Mapping

The construction of genetic maps is a prerequisite for MAS. Genetic mapping is currently being developed for more than 25 forest species with the main objective of making maps denser by positioning a larger number of markers. Genera and species include:

- *Eucalyptus*, including *E. camaldulensis, E. globulus, E. grandis, E. nitens, E. tereticornis* and *E. urophylla*;
- *Populus* in particular *P. deltoides, P. nigra* and *P. trichocarpa*;
- *Picea* including: *P. abies, P. glauca* and *P. mariana*;
- *Pinus* including *P. brutia, P. edulis, P. elliottii, P. palustris P. pinaster, P. radiata, P. strobus, P. sylvestris* and *P. thunbergii*;
- other species: *Acacia mangium, Cryptomeria japonica, Larix leptolepis* [=*L. kaempferi*], *Pseudotsuga menziesii, Quercus robur* and *Taxus brevifolia*.

Several plant functional genomics programmes have undertaken linkage map construction, and significant developments in this area have been made with some forest trees.

The search for linkages between coding regions and traits linked to wood quality has allowed some gene candidates to be identified (especially in lignin synthesis pathways) although they are not yet used in routine selection programmes. There are noteworthy EST bank development programmes underway for several species in several genera, including *Pinus*, *Populus*, *Eucalyptus* and *Betula*.

The gene candidate approach has been used in a dozen species of *Pinus*, at least five *Eucalyptus* spp. including hybrids and in three *Picea* spp.. More recently, research work has been undertaken on other traits such as water stress tolerance. Proteomic studies have been initiated on wood properties for a limited number of species: *Pinus pinaster*, *Eucalyptus gunnii*, *Populus* spp. and *Picea sitchensis*.

Micropropagation

In vitro vegetative propagation or micropropagation is aimed at cloning superior individuals or at 'bulk' (in mixture) propagating new genotypes with high genetic potential but available in limited quantities (such as materials obtained by controlled pollination). Tree species can be micropropagated by microcuttings or by somatic embryogenesis.

Micropropagation Methods

Micropropagation by microcuttings consists of mass producing vegetative copies of desired genotypes by either axillary or adventitious budding. In the latter case, differentiated cells, usually from superficial tissues, must undergo a de-differentiating process before new shoot formation is initiated. Ultimately, production of independent and self-sufficient individuals is completely dependent on the *de novo* formation of an adventitious root system. Somatic embryogenesis, or production of embryos from somatic cells, is in fact a cloning technique, as opposed to zygotic embryogenesis in which germinal cells give rise to seedlings that are all genetically different.

The process of somatic embryogenesis derives usually from callus formation induced by applying cytokinic or auxinic exogenous growth regulators to very juvenile plant tissues. In the most favourable situations, some undifferentiated cells of these calli can evolve into somatic embryos characterized, similarly to zygotic embryos, by a shoot–root bipolar structure. This basically distinguishes somatic embryogenesis from microcuttings consisting first of a shoot from which an adventitous root must subsequently develop.

The majority of GM trees is likely to be used in the form of clonal materials and will need to be vegetatively propagated. In the case of conifers, somatic embryogenesis, especially when derived from a single cell, seems the most suitable regeneration and propagation technique. In broad-leaved species, vegetative propagation of GM materials is likely to use a combination of micropropagation and rooted stem cuttings, at least in the beginning.

Database Analysis

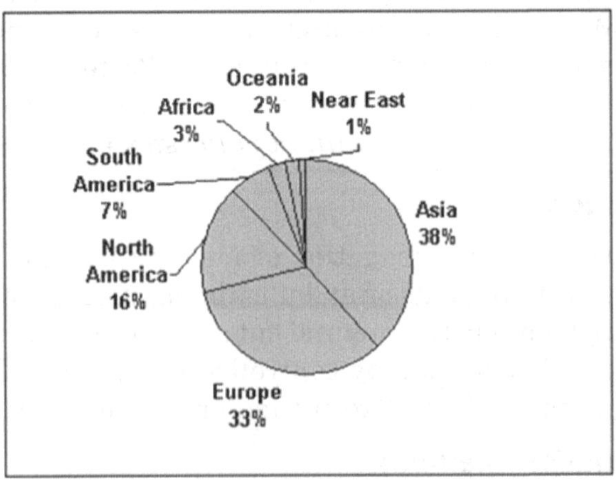

Figure: *Distribution of micropropagation activities by region of activity.*

Asia accounts for 38 percent of documented activities in forest tree micropropagation, followed by Europe (33 percent), North America (16 percent), South America (7 percent), Africa (3 percent), Oceania (2 percent) and the Near East (one percent) (Figure 13A). Sixty-four countries active in this field have been identified. A closer look shows that species from Asia (27 percent) and Europe (21 percent) predominate (Figure 13B), which may suggest that the activities in Asia are concentrated on indigenous species, while *Eucalyptus* spp. are micropropagated in all continents. Species originating from South America and Africa are almost absent from the data set. Micropropagation activities seem to take place mainly in countries with significant tree planting programmes.

Most micropropagation research activities are at a very advanced stage. Germplasm and protocols are likely to be available for large-scale deployment soon, if not already operational, for species and genera including:

- *Anogeissus* in India;
- *Acacia mangium* and *A. mangium* × *A. auriculiformis* in Malaysia, Indonesia and Vietnam (O. Monteuuis, personal observation);
- *Eucalyptus* in Vietnam, India (O. Monteuuis, personal observation), Australia and South America;
- *Pinus* in Canada and New Zealand, and *Pinus taeda* in the USA (Levis W. Handley, personal communication);
- *Tectona grandis* in Brazil, Thailand and Indonesia (O. Monteuuis personal observation) and Malaysia (Goh *et al.* in press).

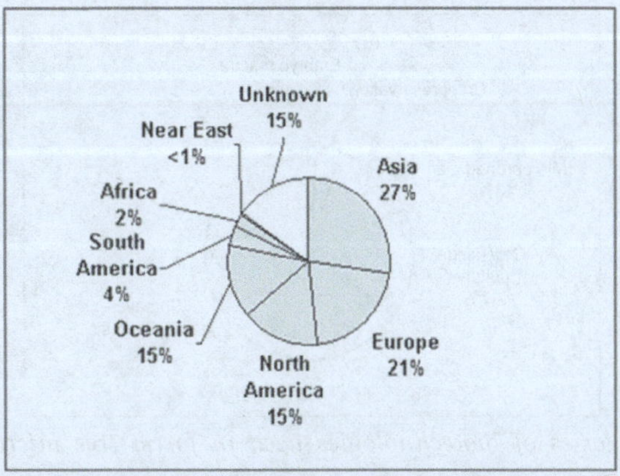

Figure: *Distribution of micropropagation activities by region of origin of the species studied.*

Micropropagation techniques have been tested on at least 82 forest tree genera (many of them of proven or potential interest in forest plantations). Research work has been devoted to *Pinus* (13 percent), *Picea* (13 percent), *Eucalyptus* (11 percent), *Acacia* (7 percent), *Quercus* (6 percent), and *Populus*, *Larix* and *Tectona grandis* (4 percent each) (Figure 14).

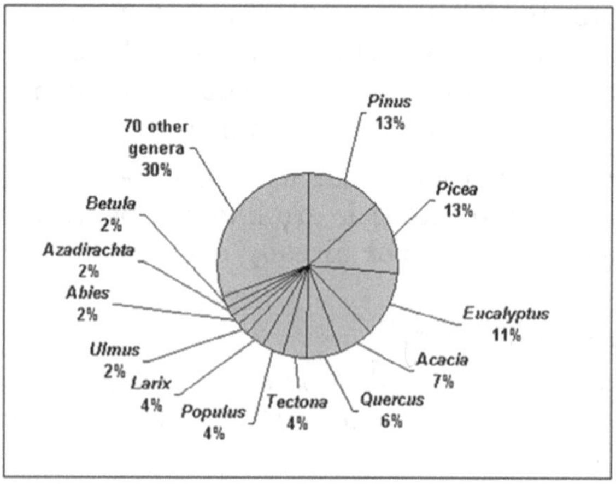

Figure: *Distribution of micropropagation activities by genus.*

The first five genera account for 50 percent of all activities carried out in forest tree micropropagation. Somatic embryogenesis accounts for 65 percent of activities in micropropagation followed by cell/tissue culture (17 percent), micropropagation by microcuttings in *in vitro* conditions (13 percent), cryopreservation/conservation (4 percent) and embryo rescue (one percent) (Figure 15).

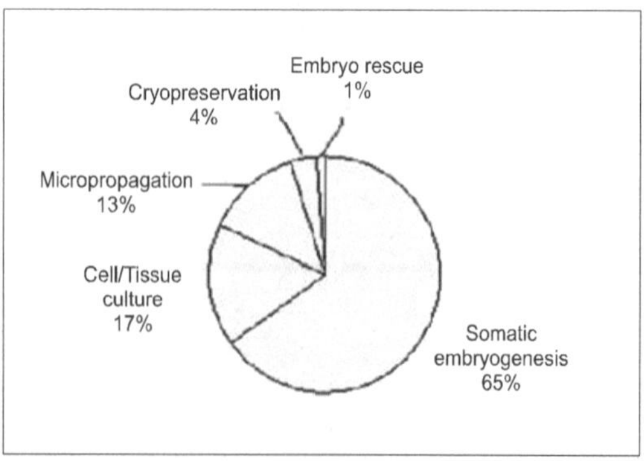

Figure: *Categories of biotechnologies used in forest tree micropropagation.*

Micropropagation by microcuttings is carried out on more than twenty species including:

- *Populus alba, P. deltoides, P. tremula* and *Populus* hybrids in Germany and India, Spain and Lithuania;

- *Eucalyptus camaldulensis, E. globulus, E. grandis, E. nitens, E. tereticornis* and *E. urophylla* in South Africa, Spain and Portugal, India, Vietnam and Thailand, and Australia;
- *Acacia mangium, A. melanoxylon, A. mangium* × *A. auriculiformis* in Malaysia and South Africa;
- *Tectona grandis* in India, Vietnam, Brazil and Indonesia (O. Monteuuis personal observation), Thailand (Kjaer *et al.* 2000), Costa Rica (Schmincke 2000), Malaysia and Australia (Monteuuis, personal observation).;
- *Larix* in Canada and France;
- *Pinus radiata, P. taeda* and *P. pinaster* in France, New Zealand and the United States;
- *Pseudotsuga menziesii* in the United States;
- *Sequoia sempervirens* and *Sequoiadendron giganteum* in France and Germany;
- *Anogeissus latifolia* and *A. pendula* in India;
- *Betula pendula* in Norway and Finland;
- *Paulownia fortunei* in Australia (O. Monteuuis personal observation);
- *Platanus acerifolia* in China;
- other species include *Gmelina arborea, Artocarpus chaplasha, A. heterophyllus, Azadirachta indica* and *Elaeocarpus robustus* in Bangladesh.

The advantages of somatic embryogenesis in comparison with micropropagation by microcuttings, especially with regard to multiplication rate and genetic modification applications, explain the major research investments in the technique. However, there are still serious obstacles to large-scale operational application of somatic embryogenesis to forest trees, for example:

- Only some species and, within these species, only some genotypes can produce somatic embryos.
- Success has been obtained, with few exceptions, mainly with juvenile tissues coming for instance from immature zygotic embryos.
- There are risks that somaclonal variation may decrease the value of the genotypes produced by somatic embryogenesis, resulting in a considerable waste of time, material and money.

True-to-typeness, particularly, may remain a problem for certain genotypes, and efforts are still needed for optimizing this technique to make it more reliable, especially when using mature selected genotypes.

Although activities on somatic embryogenesis have been reported in all regions of the world, it seems that countries with conifer planting programmes show a stronger interest in this micropropagation technique:

- *Larix* in Canada (and France);
- *Picea abies, P. glauca, P. mariana* and *P. sitchensis* in Canada, France and Ireland;
- *Pinus banksiana, P. patula, P. radiata, P. strobus* and *P. taeda* in Canada, South Africa, New Zealand, the United States and Chile;
- *Pseudotsuga menziesii* in the United States and Canada;
- *Eucalyptus globulus, E. grandis* and *E. dunnii*;
- *Gmelina arborea, Artocarpus chaplasha, A. heterophyllus, Azadirachta indica* and *Elaeocarpus robustus* in Bangladesh.

Investments in vegetatively micropropagated species of commercial interest have resulted in successful large-scale applications in several countries in Europe, North America, Asia and Oceania. Unfortunately, there is a lack of basic information on areas planted with forest vitroplants.

Genetic Modification In The Forestry Sector

The first regeneration of a genetically modified (GM) forest tree was achieved in 1986 in *Populus*. Since then, the genus has become a model for genetic modification and related tree biotechnology studies.

The first attempt to genetically modify a conifer (*Larix*) was reported in 1991. Introducing targeted genes into the genome of a forest tree is a way to obtain GM plants. It is also a basic research tool for a better understanding of gene functioning in woody plants.

Review of Objectives and Methodologies

Transformation methods consist of inserting into the genome of the host plant a mutated gene or a gene from another organism either by microinjection/projection (direct transformation), or through a vector such as *Agrobacterium tumefaciens* (indirect transformation).

Most successful work on genetic modification of forest trees species genomes so far has been obtained by using juvenile material, often from explants produced from early germinations tissues which have much higher regeneration capacities than older material. Successful genome modification reports of adult selected plant material are very rare, except in poplars precisely because of its greater capacity to regenerate.

Herbicide Resistance

A first method has consisted in introducing a mutated version of the gene encoding the enzyme target for various herbicides: glyphosate for *Populus alba* × *P. grandidentata, P. trichocarpa* × *P. deltoides, Eucalyptus grandis, Larix decidua* and *Pinus radiata* (the herbicide blocks the synthesis of tryptophan, tyrosine and phenylalanine), or chlorosulfuron for *Populus tremula* and *Pinus radiata* (the herbicide blocks the synthesis of leucine, isoleucine and valine). A second and more frequently used strategy consists of introducing a microbial gene encoding an enzyme for the detoxification of the herbicide, and has been applied to *Populus alba, P. alba* × *P. tremula, P. tremula* × *P. alba, P. trichocarpa* × *P. deltoides* and *Eucalyptus camaldulensis.*

Insect Resistance

The first strategy consists of using microprojectiles to insert a gene encoding an endotoxin which binds to the receptors in the intestine of Lepidoptera, Coleoptera and Diptera, lysing the organ and killing the insect. This was done in *Populus alba* × *P. grandidentata, P. tremula* × *P. tremuloides* and *Picea glauca.* GM *Populus* trees and *Pinus radiata* expressing the *Bacillus thuringiensis* endotoxin '*Bt*' have been obtained. A second transformation method is based on the introduction of a gene coding for a protease inhibitor that modifies insect digestion, causing the death of the pest. Studies used potato gene *pin*2, a protease inhibitor introduced into *P. alba* × *P. grandidentata* through *A. tumefaciens*, and the gene of a rice protease inhibitor introduced into *P. tremula* × *P. tremuloides.* A third approach focuses on simultaneous modification of two genes for enhanced resistance to insects. This was achieved in *Liquidambar styraciflua*, combining a peroxidase anionic enzyme gene involved in cell growth and wall development with a '*Bt*' gene.

Flowering Modification

In order to produce sterile trees and prevent possible dispersal of transgenic pollen in the environment, an approach based on genetic ablation has been tested on poplar. This technique, which consists of expressing a cytotoxic gene under the control of a very specific poplar floral promoter, resulted in more than 90 percent of transformed lines lacking floral structures. Other approaches, based on the suppression of key flowering genes, are being tested.

Quantitative and Qualitative Modification of Lignin

Modification of lignin composition or content is being actively pursued because of the expected financial gains from pulp processing improvements.

Lignins, which enhance cell wall mechanical properties and hardness, are difficult to process and are a significant limitation in processing wood into paper pulp by chemical treatment. Genetic transformation to modify lignin characteristics is a key research feature on species used in the paper industry. The aim is to regulate the activity of key enzymes involved in the lignin biosynthesis pathway. Active on-going research targets the effects of lignin biosynthesis in *Populus* on soil carbon transformation and storage.

Other Traits

Other traits that have been considered for genetic modification include: resistance to diseases in *Populus*, *Betula pendula*, *Alnus* and *Juglans*.; bioremediation in *Populus*; enhanced growth through higher efficiency of nitrogen assimilation in *Picea* and *Populus*, and modification of gibberellin synthesis in *Populus*.

Data Set Analysis

Genetic modification activities represent 520 entries in the data set, or 19 percent of all biotechnology activities recorded worldwide. Research is reported in 35 countries and on 29 tree genera. North America is the lead region, accounting for 48 percent of genetic modification activities on forest tree species, followed by Europe (32 percent), Asia (14 percent), Oceania (5 percent) and Africa and South America (less than one percent each) (Figure 16), while no activity is reported in the Near East.

Genetic modification research at the country level is led by the United States (219 entries in the data set, or 42 percent of the world's recorded activities on GM forest trees) (Figure 17). Studies carried out by, or in cooperation with, agencies or companies from developed countries have been identified in Brazil, Chile, China, Indonesia, South Africa and Uruguay.

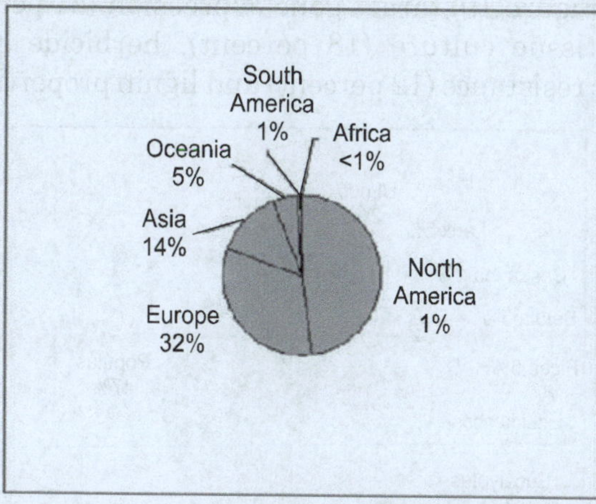

***Figure:** Distribution of data set entries on genetic modification by region of the world.*

Populus accounts for 47 percent (Figure 18) of data set entries, followed by *Pinus*. (19 percent), *Eucalyptus* (7 percent) and *Liquidambar* and *Picea* (5 percent each). The pedigree of a number of hybrid poplars used in China has not been ascertained. Several countries have already established GM tree field trials.

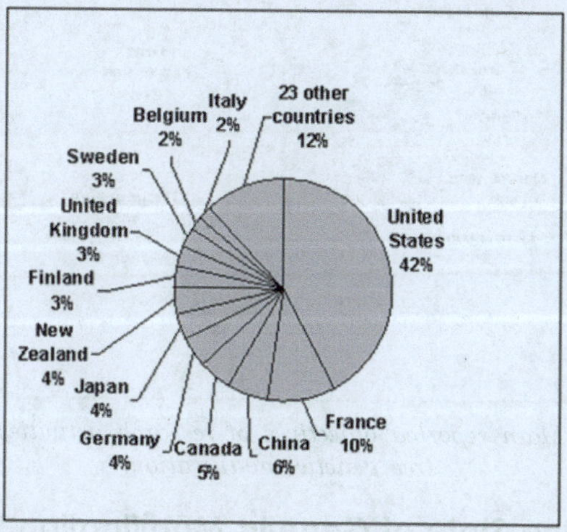

***Figure:** Distribution of reported forest tree genetic modification research activities by country (only those with more than five such activities in the data set are included).*

Interestingly, while the first GM forest trees were developed some fifteen years ago, about three-quarters of the total number of species recorded in the dataset have been genetically transformed during the past five years. The main studies underpinning genetic modification (Figure 19) target gene expression (21 percent), *in vitro* regeneration tissue culture (18 percent), herbicide resistance (13 percent), biotic resistance (12 percent) and lignin properties (9 percent).

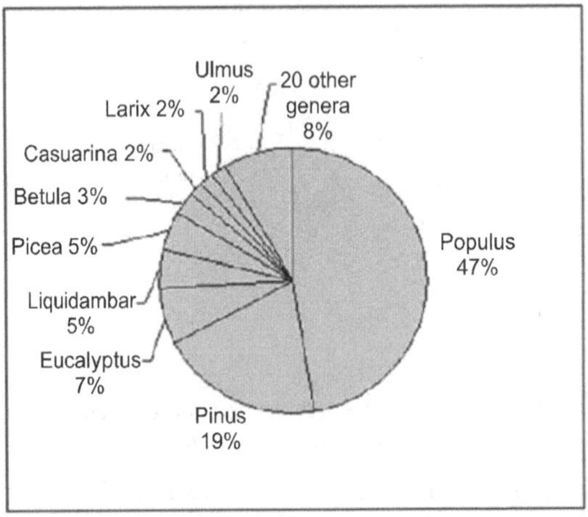

Figure: *Distribution of reported forest tree genetic modification research activities by genus*

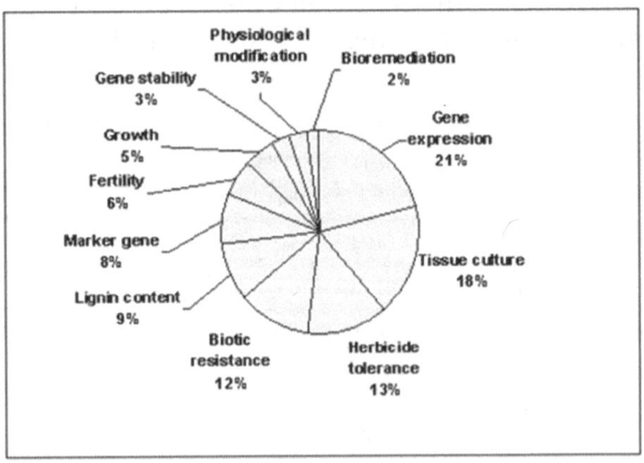

Figure: *Main reported objectives of research activities in forest tree genetic modification.*

Summary of the State of Genetic Modification in the Forestry Sector

In vitro regeneration of transformed plants is still a technical limitation for many species and genotypes. Genetically transformed lines have been regenerated for *Populus*, *Eucalyptus*, *Pinus* (including *P. taeda*) and *Picea*. Technical processes involved in regeneration are not always well controlled or even studied; especially in mature tissues where the situation is the most critical. No GM tree is reported to be commercialized in international databases to date. Insertion of genes controlling traits of interest in a given forest species is limited by insufficient knowledge of the molecular control of these traits.

Most of the traits important in wood production, such as growth rate, adaptability, and stem and wood quality, are under polygenic control, and therefore depend on more than a single gene. Biosafety regulations governing GM trees are often a concern due to the long period of time necessary to monitor environmental effects (including gene stability, pollen and seed dispersal, and impact on other elements of the ecosystem).

Most existing regulations are both stringent and involve high development costs compared to annual crops. Collaborative genomics programmes such as those carried out on *Populus, Pinus* and *Eucalyptus*, with the on-going characterization of genes of interest (including those coding for flowering, and lignin and cellulose biosynthesis), and the identification of specific promoters of transgenic expression, are expected to boost advances in genetic modification of forest trees. Studies to assess the stability of GM tree growth under abiotic stress, and their environmental impact, have been initiated.

Conclusions: Status and Trends of Biotechnology Research and Applications in Forestry

Unlike work carried out on crop or farm animal species, the domestication of forest trees is, with few exceptions, very recent. However, within a few decades, knowledge transfer from agriculture to forestry has been so fast that genetic knowledge of a few tree model genera, such as poplar (*Populus*), pine (*Pinus*) and *Eucalyptus*, is about to become as important, in terms of complexity and quality, as the genetic knowledge of the main crop species.

Modern biotechnologies have potentially outstanding applications in the forestry sector, mainly because of the potential genetic gains they could confer. Expected gains for tree breeders include new genetic pools and significant reduction in tree selection time.

Owing to technical difficulties, high costs, and time needed to evaluate forest species, tree breeding programmes have long been limited to the mass selection of the best individuals. Between the 1970s and 1990s, development of horticultural techniques such as grafting, propagation by cuttings and controlled pollination, as well as information technology development, contributed to the implementation of sophisticated and efficient tree improvement activities. Long-term conventional tree breeding programmes and large-scale clonal propagation techniques have resulted in, and will continue to provide, significant productivity gains, especially for fast-growing species such as *Eucalyptus* and *Populus*. While the objectives of tree breeding are clearly defined and address end-users' (tree growers') requirements (in terms of yield, wood quality, and biotic and abiotic resistance), the situation in biodiversity conservation is less clear. Several issues (including knowledge of species biology, ecosystem conservation, restoration of endangered species, and decision-making tools for ecosystem or forest management) are common to biological diversity and forestry.

However, in spite of a considerable number of studies carried out on hundreds of forest tree species, there are still very few reported biotechnology applications. Many countries are currently engaged in biotechnology-supported forest biodiversity conservation programmes. Nonetheless, there are only a few reported direct impacts at policy or technical levels.

The impact of biotechnology on genetic studies has allowed the use of cutting-edge methodologies with forest tree species in spite of the late domestication process compared with crop species (Sedjo 2003). The sequencing of the *Populus* genome has significant implications for forestry. *Eucalyptus* gene sequencing, which was reported during a IUFRO (International Union of Forest Research Organizations) Tree Biotechnology meeting in 2003, has since been pursued by the Eucalypt Genome Initiative, the Eucalyptus Genome Sequencing Project Consortium and the Brazilian project Genolyptus.

The Dendrome Project is a collection of forest tree genome databases and other forest genetic information resources, with a special emphasis on conifers and *Pinus*. It is in genome sequencing that research seems currently most active.

The development of large-scale vegetative propagation techniques, essentially based on micropropagation or somatic embryogenesis, makes it theoretically possible to deploy superior planting materials rapidly and effectively.

This is already the case for some coniferous and broad-leaved tree species including *Eucalyptus* and teak (*Tectona grandis*). Micropropagation techniques are also necessary to regenerate GM plants. Genetic modification has been applied to forest trees mainly during the last two decades, at the experimental stage in the laboratory, to alter various traits such as herbicide, metal, salt, or insect resistance or cold tolerance. The rationale for producing trees with reduced lignin content for the pulp and paper industry is generally based on the expected environmental and economic benefits. Equally if not more important, genetic modification is often considered as a tool to improve knowledge of tree biology and functioning, with significant potential applications; for example, for studying cell wall properties and wood formation.

Since genes regulating secondary tissue and lignin formation in the annual plant *Arabidopsis thaliana* were identified, research on the genetic basis of wood quality in forest trees has significantly increased (mainly in *Eucalyptus, Pinus* and *Populus*). With the exception of micropropagation, for which some applications are reported, most public domain information relates to research activities. Field applications appear extremely limited, according to the literature.

Nevertheless, and in spite of unclear economic rationale, the use of biotechnology in forestry research has increased and provided new knowledge on tree biology and functioning. It also appears that forest biotechnology is increasingly funded by the public sector in academic research, while the private sector seems to focus investments more on specific traits of commercial interest such as wood quality.

Commercial deployment of new biotechnology-based forest tree varieties of a limited number of taxa (*Eucalyptus*, *Pinus taeda*, *P. radiata*, *P. pinaster*, *Populus*) can be expected in the near future.

The impact of biotechnology applications on global wood supply may, however, take longer to materialize (not before 2020, according to Seppälä 2003). In this context, genetic modification appears to be the most significant and controversial sector of forest biotechnology. It is noteworthy that in some countries the terms 'biotechnology' and 'genetic modification' are used as synonymously, adding to the confusion.

During the last decade, several collaborative groups or consortia have been created by public or private partners for applying specific forest biotechnologies in a number of countries, including:

- ArborGen, created in 1999, with the aim of enabling research on genetic modification, has been working on herbicide resistance, growth and fibre quality for paper pulp in *Eucalyptus, Pinus radiata*, *Populus*, *Pinus taeda* and *Liquidambar* sp.

- GenForSA, created in 1999 with the main objective of modifying and testing disease resistance of *Pinus radiata*, while improving wood quality and formation.
- Monfori Nusantra Indonesia, created in 1996, has been involved in mass production of tissue cultures of *Tectona grandis, Acacia* and *Eucalyptus* for field trial establishment and commercialization.
- The Oregon State University's Tree Genetic Engineering Research Cooperative (TGERC) has carried out research on the use of GM poplar in plantations.
- The Poplar Molecular Genetics Cooperative (PMGC) has been established with the aim of improving knowledge of genetic and molecular mechanisms responsible for productivity and quality trait variations in hybrid poplars.
- The North Carolina State University Forest Biotechnology Industrial Research Consortium (FORBIRC) mission is to integrate genome technology, metabolic engineering, traditional tree breeding and wood and paper science into a research organization directed to the creation of superior wood as a raw material and as a product.

The interest induced by the first genetic modifications of tree species at the end of the 1990s has since declined. Regulatory frameworks have been established in many countries for GM organism (GMO) field experiment or commercial deployment, with increased testing costs. In a number of countries, hostile reaction from the public or/and environmental groups has also limited deployment of GM trees. Some authors have questioned the profitability or return on investment of MAS techniques, at least in the short term and the medium term.

In Europe, forestry research priorities now focus on stronger synergies between biotechnology and conventional selection and breeding programmes mainly for wood improvement. The aim is to fill the gap between forest trees and agricultural crops in terms of the level of domestication. As a result, tree improvement tends to use genomic and proteomic approaches for achieving trait selection more rapidly. This is done by taking advantage of the advances obtained in model plants such as *Arabidopsis*, and of the considerable increase of information exchange resulting from the latest bioinformatics progress. More and more effort is devoted to research aimed at better understanding biotic and abiotic stress responses. Significant investments have been made in genomics research on *Populus* and *Pinus* within the framework of the Poplar Genomics Initiative and the Pine Mapping Project.

In the tropics, wood production in volume per hectare is significantly higher than in the temperate zone. Several large-scale companies have recently invested in tree planting in tropical and subtropical areas. Some paper companies tend to fund their own research activities and keep them confidential.

More emphasis seems to be put in genomics and proteomics, and particularly sequencing, thanks to developments on *Arabidopsis thaliana*, *Oryza* and *Populus*.

At the same time, there is increasing information coming from genome banks on wood quality, growth and stress resistance. Good examples of EST development include *Eucalyptus* in the framework of the Genolyptus project in Brazil, *Populus*, *Pinus taeda*, and *P. pinaster* and *Quercus*, although applications for practical field selection are very unlikely to be operational for at least a decade.

Some predict that the next few years are more likely to see significant field applications of forest genomics research than of genetic transformation research, whose application is seriously hindered by strong environmental opposition and stringent regulations.

Developments in agricultural crops are very meaningful is this respect. Genomics and proteomics outcomes are expected to boost conventional tree breeding and improve the efficiency of existing programmes.

The production costs of a new forest tree variety using conventional selection (well-known species, with a short juvenile phase) is roughly estimated at about US$400 000 and requires 15 to 20 years to be developed (Fenning and Gershenzon 2002).

The calculation does not include the investment required for the selection of the suitable species. On the other hand, genomics, supported by bioinformatics in the framework of MAS, requires substantial equipment and running budget (as an indication, a supportive biotechnological platform costs several million dollars).

In turn, such biotechnologies are expected to enable breeders to access more accurate and numerous traits of interest for greater improvements, while reducing the new variety production costs thanks to bioinformatics.

As regards developing countries and countries with economies in transition, there are few references available on their involvement in forestry biotechnology. The limited literature mainly refers to micropropagation in Vietnam, Malaysia, Indonesia and India. Malaysia has a reported strong oil palm molecular biology programme, including genetic modification.

However, some emerging countries with advanced financial, institutional and human capacities (including Brazil, India, and China) have made significant breakthroughs in advanced forest biotechnology. Brazil has been actively involved in research on *Eucalyptus* through the Genolyptus project that constitutes at the moment the cutting edge in research on the genus worldwide. China is very active in poplar genetic modification.

It can be reasonably assumed that after a period of time, results from research in technically advanced countries will benefit an increasing number of developing countries and countries in transition. African countries, with the exception of Kenya and South Africa, are far less advanced in forest biotechnology. However, notable development of agricultural crop biotechnology in countries like Nigeria, Uganda, Côte d'Ivoire, Morocco and Algeria (particularly through AAB) and even in the forestry sector (ISAAA in Kenya), constitute positive indicators suggesting that the situation is improving. Adoption of well-targeted forest biotechnology, possibly for drought-resistance, seems a crucial issue for the future of such activities.

Micropropagation has for the past several years been the most striking example of forest biotechnology applications, at least in terms of numbers of plants produced and acclimatized. Applications have been reported mainly in Asian countries, although there is still a lack of reliable information as far as operational field planting is concerned.

Forestry Sector Conference

Plant biotechnology is a field of scientific research in which rapid advances have been made in recent years and which appears to have much potential for further development. Numerous opportunities for using biotechnology in plant breeding have been identified, some of which might be appropriate for the improvement of crops in developing countries, as discussed in the crop sector conference. In this conference the focus is on forest trees and currently available biotechnologies and their application in the forestry sector are discussed with reference to their potential use in developing countries today. Please note that for the purposes of this conference, the term "forestry sector" specifically excludes fruit orchards. Most forest tree species are characterized by inherently high levels of variation and extensive natural ranges. This high level of genetic variation needs to be maintained to ensure present-day and future adaptability to changing environmental conditions. It is also needed to maintain options and potential for improvement to meet changing end-use requirements. Forests provide a wide range of goods and services such as timber, fibre, fuelwood, food, fodder, gum, resins, medicines, pharmaceutical products and environmental stabilization. Similar goods and services are often provided by a wide range of genera and tree species. Despite the availability of a large number of forest tree species, less that 500 have been systematically tested for their present-day utility for human beings and less than 40 species are included in intensive selection and breeding programmes.

Selection in breeding populations with a broad genetic base is the most common approach to forest tree improvement. Although demand for wood is the driving force in the development of large-scale forest plantations, several selection and breeding programmes aim at enhancing other goods and environmental services provided by forest trees and shrubs.

Since most forest tree species are characterized by long generation intervals and a generally long juvenile phase before flowering, much time is needed before assessment of important traits can be carried out. For example, if wood quality is of interest in breeding for timber or fuelwood, selection can only be carried out after trees have reached a certain size which, in some cases, can require decades. The above factors are limitations to rapid improvement and only a maximum of three or four generations of breeding have been completed in a few forest tree species to date (*Eucalyptus grandis* and some pine species).

Description of Biotechnologies in the Forestry Sector

Biotechnologies Based on Molecular Markers

Reliable information on the distribution of genetic variation is a prerequisite for sound selection, breeding and conservation programmes in forest trees. Genetic variation of a species or population can be assessed by measuring morphological and quantitative characters in the field or by studying molecular markers in the laboratory. A combination of the two methods is required for reliable results. Molecular markers can be used for:

Quantification of Genetic Diversity

The use of molecular markers for the determination of the extent of variation at the genetic level, within and between populations, is of value in guiding genetic conservation activities, which are aimed at maintaining genetic diversity with respect to traits of both known and unknown importance, and in the development of breeding populations for specific end uses.

It should be noted that studies on genetic diversity based on molecular markers must be interpreted with caution, due to frequently low correlations with patterns of variation for adaptive traits, which are of major importance in forestry.

Genotype Verification

Molecular markers have been widely used for identification of genotypes and applied in taxonomic studies, biological studies and 'genetic fingerprinting'. Good taxonomy is fundamental to conservation and tree improvement programmes and to programmes involving hybridization between species.

The use of molecular markers has revolutionized studies of mating systems, pollen movement and seed dispersal. Results of such biological studies are of considerable practical significance to advanced tree improvement programmes, specifically in population sampling, design and management of seed orchards (i.e. orchards consisting of clones or seedlings from selected trees and cultured for early and abundant production of seeds for reforestation), estimation of pollen contamination and development of controlled pollination methods. Germplasm identification, through 'genetic fingerprinting', has been used in advanced breeding programmes which rely on controlled crosses or in which the correct identification of clones for large-scale propagation programmes is essential.

Gene Mapping and MAS

Genetic linkage maps can be used to locate genes affecting quantitative traits of economic importance. Quantitative traits, such as wood yield, wood quality or pulp yield, are usually controlled by many genes, termed QTLs. By using molecular markers closely linked to, or located within, one or more QTL, information at the DNA-level can be used for early selection. The potential benefits of MAS are greatest for traits that are difficult, time-consuming or expensive to measure (for example, wood quality traits or pulp yield). Mapping and

MAS tend to be used only in species of high economic value and have most potential in clonal breeding programmes, where additional genetic gains can be rapidly multiplied.

Biotechnologies Based on Vegetative Propagation

Strategies supporting large-scale utilization of genetic material with a narrow genetic base must be appropriately integrated into tree improvement programmes. Vegetative propagation within such programmes allows for a fast release of new materials and for appropriate matching of clones to different local conditions. It also allows continued cultivation of given clones and to efficiently change the mixture of clones used in a given programme. Vegetative propagation also supports other currently available biotechnologies (*in vitro* storage and cryopreservation; *in vitro* selection).

In Vitro Storage and Cryopreservation

In vitro storage refers to the storage of germplasm in aseptic culture under laboratory conditions, while cryopreservation refers to the storage of cells, tissues, seeds, etc. at temperatures of liquid nitrogen (-196ºC). The two techniques do not seem to be widely used in genetic conservation activities for forest trees, but they may serve as back-up strategies for species with seed storage problems.

In Vitro Selection

In vitro selection refers to the selection of germplasm based on test results using tissue culture under laboratory conditions. Many recent publications for crop plants have reported useful correlations between *in vitro* responses and the expression of desirable field traits, most commonly disease resistance.

Positive results are available also for tolerance to herbicides, metals, salt and low temperatures. For the selection criteria of major general importance in forest trees (in particular vigour, stem form and wood quality), poor correlations with field responses will limit the usefulness of *in vitro* selection. However, *in vitro* selection may be of possible interest in forestry programmes for screening disease resistance and tolerance to salt, frost and drought.

Micropropagation

For crop and horticultural species, micropropagation (*in vitro* vegetative propagation of plants) is now the basis of a large commercial industry involving hundreds of laboratories around the world.

Successful protocols now exist for a large number of forest tree species and the number of species for which successful use of somatic embryogenesis has been reported is increasing (somatic embryogenesis is a step in micropropagation where somatic cells are differentiated into somatic embryos). So, in the future, it is likely that micropropagation in the forestry sector will become commercially more important. Compared to vegetative propagation through cuttings, the higher multiplication rates available through micropropagation seem to offer a quicker capture of genetic gains obtained in clonal forestry strategies.

One major factor impeding early application of micropropagation in many large-scale forest plantation programmes, is that breeding and selection of desired clones are not sufficiently advanced for clonal forestry to be contemplated. Current high costs will also be an impediment to the direct use of micropropagation in many programmes. Technologies resembling those used commercially in horticulture are most likely to be affordable for a limited number of high-value forest tree species, particularly those for which propagation by cuttings is difficult. Micropropagation is unlikely to be used for the production of planting stock of non-industrial forest tree species.

Genetic Modification of Forest Trees

GMOs are defined as organisms that have been modified by the application of recombinant DNA technology (where DNA from one organism is transferred to another organism). The term "transgenic trees" is also used for GM trees, where a foreign gene (a transgene) is incorporated into the tree genome. One of the first reported trials with GM forest trees was initiated in Belgium in 1988 using poplars. A study carried out in 1999 indicated that, since then, there have been more than 100 reported trials, involving at least 24 tree species-most of which are timber-producing species.

The majority of the field trials were carried out in the USA and Canada. Whereas it is estimated that roughly 40 million hectares of transgenic agricultural crops were grown commercially in 1999 (ISAAA figures), there is no reported commercial-scale production of transgenic trees. Information on field trials of GM trees has been published by the OECD and the World Wide Fund for Nature (1999). Traits for which genetic modification can realistically be contemplated in the near future include insect and virus resistance, herbicide tolerance and lignin content.

However, insertion of any gene into a tree species with expected functional results will be a substantial undertaking and insertion of enough genes to confer e.g. long-term insect resistance in a perennial species even more so. Virus and insect resistance, in particular, are of major significance for crop plants. By contrast, these traits are not the most important in breeding programmes of forest tree species (poplars being an exception). Reduction of lignin is a valuable objective for species producing pulp for the paper industry; work on this aspect is underway in aspen.

A major technical factor limiting the application of genetic modification to forest trees, is the current low level of knowledge regarding the molecular control of traits which are of most interest, notably those relating to growth and stem and wood quality.

Genetic modification of these traits remains a distant prospect. Investments in genetic modification technologies should be weighed against the possibilities of exploiting the large amounts of genetic variation, which are generally untapped, available within any single species in nature. Biosafety aspects of GM trees need careful consideration because of the long generation time of trees, their important role in ecosystem functioning and the potential for long distance dispersal of pollen and seed.

Forestry in Developing Countries

Forests cover approximately 30 percent of the world's total land area. They are the source of vital commodities, including raw materials and food and are essential for maintaining agricultural productivity and the environmental well-being of the planet as a whole. They protect soil and water and buffer the effects of wind and rain, thus helping to decrease soil erosion and they are an important sink for carbon dioxide. Forests are also among the most important repositories of biological diversity. Roughly 500 million rural people live in, or close to, forests.

Most communities use a variety of forest products, particularly those in developing countries. Plant stems, tubers and fruits provide additional food during hungry seasons or when crops fail; wild animals are harvested for meat and hides; and the forests provide fuelwood, fodder for livestock, medicines and other products and services. The most important trend in forestry in developing countries is the progressive reduction in the area of forests due to changes in land use. Another important trend, evident at a global level, is increasing forest degradation through unmanaged use. When forests are degraded, their productive functions and their capacity as regulators of the environment are reduced, increasing flood and erosion hazards, reducing soil fertility and contributing to the loss of forest products and overall loss of biological diversity.

While forests are being lost, there is growing demand both for environmental services and for wood and wood products which they provide. A forecast by FAO predicts that wood demand is expected to increase by 25 percent from 1996 to 2010.

This demand will, increasingly, have to be met by forest plantations, and with decreasing land areas available for forestry, plantation methods will have to be increasingly intensive. This will necessitate better tree improvement programmes in which biotechnology may play a role.

6

Agricultural Productivity for Sustainable Food Security through Forestry

Ensuring food for all, today and in generations to come, is one of the greatest challenges facing the world community. Food security is defined as the ability of people to meet their required level of food consumption at all times; it is considered by many to be a basic human right. However, about 1.1 thousand million people in low-income, food-deficit developing countries cannot meet such basic needs.

Among them, more than 800 million live in rural areas, depending directly on agriculture for their food supply, employment and income. Therefore, boosting the rural economy, particularly through increased agricultural production, is one of the chief means of alleviating poverty and increasing food security.

Food security consists of three major components: availability (associated with production and trade); accessibility (associated with income and wealth); and utilization (associated with health and nutrition) (Asenso-Okyere, Benneh and Tims, 1997). While there seems to be a consensus among analysts that current global food production is adequate to avoid widespread famine and malnutrition, the overall positive trends disguise the disparities in production and distribution of food between regions.

The disparities are described by Rosegrant, Agcaoili-Sombilla and Perez (1995) as a two-tiered system of food security, in which rich and rapidly growing economies enjoy abundant, affordable food supplies, while poor, slow-growing countries suffer from food scarcity and malnutrition. This means that the food security problem is, in the main, not one of shortage but of imbalance and distribution. Another aspect of food security is sustainability.

The concerns are: "can food production continue to keep up with demand in generations to come?" and "is the prosperity of the current generation at the expense of the future?" Some analysts believe that the rapid growth in agricultural production in the last few decades has occurred at great environmental cost (Anderson, 1994).

That is, over-exploitation has resulted in natural resources being depleted and the environment being damaged. Indeed, the greater intensity of use of land and water resources and chemicals has created problems such as soil salinization, soil erosion, water pollution, pest resistance, etc. As a result, there are signs of declining rates of growth in yields. For example, it has been shown that the average annual growth rate of paddy rice yield in the world declined from 2.42 percent in 1974-82 to 1.78 percent in 1982-90.

The corresponding figures were 2.62 and 1.66 percent for Asia and 4 and 1.6 percent for China. Similar results were found for other crops, including wheat, maize, sorghum and other coarse grains.

Similarly, externalities from agricultural production, and related environmental or "green" issues such as climate change, preservation of wilderness areas and biodiversity, animal welfare and food safety, have received increasing attention in the discussion of agricultural policy in recent years.

These concerns suggest that sustainable food security is not only about meeting the increasing and changing demand for food now, but about protecting the environment for future generations. Whether and how this is to be achieved depends on a number of economic, social and political factors, both at the national and international levels. Socioeconomic factors with potentially significant effects on future developments in the world food situation include: population and income growth, demographic changes and urbanization on the demand side, as well as technological change and productivity growth on the supply side.

Therefore, future agricultural production and productivity growth depend on, among others, a combination of agricultural, environmental, trade and macroeconomic policies at the global level. Although food security issues are multi-faceted, the discussion here focuses on food availability and production in Asia and the Pacific as a first step towards resolving such issues. Particular attention is paid to the role of investment and agricultural productivity in meeting the challenge of sustainable food security.

Seven member countries in this region were selected for in-depth examination: Australia, the United States, China, India, Indonesia, Japan and South Korea. These countries are chosen because of their importance in the food balance both in the world and the region.

For example, China and India are predicted jointly to account for more than 30 percent of the estimated global increase in cereal demand (718 million metric tons) between 1993 and 2020 (Rosegrant and Ringler, 1997).

The estimated demand increase from China has raised concerns over China's ability to feed itself and the impact of changes in China's trade position on global food balances and prices. Indonesia is chosen for similar reasons. Together, these three countries account for nearly 70 percent and 40 percent of the population in Asia and in the world in 1996, respectively.

In addition, Australia and the United States are chosen because of their role as major food exporters to the region. Japan and South Korea, on the other hand, represent (newly) industrialized countries that are major food importers. These differences among the selected countries in the stage of economic development, resource endowments and government policy are central to identifying factors in affecting agricultural productivity growth and the role of government policy.

Production and Productivity Growth in Agriculture

When evaluating the performance of a production unit or the agricultural sector, it is common to use production (the level of output), productivity (output per unit of input) or efficiency (actual output relative to the potential output or best practices) as indicators. Although these measures are closely related, they can yield different rankings in measuring performance.

In general, productivity is the most commonly used measure, be it measured in terms of total factor productivity (TFP) or in partial terms such as labour productivity (output per labour) and yield (output per hectare) for its relative ease in calculation and interpretation. In the following sections, changes over time in output and productivity growth in different regions are compared and the causes for variations are discussed.

Agricultural Development and Input Use

Agricultural output and productivity vary greatly with the stage of economic development, resource endowments, government policy and agronomic-ecological conditions. However, there is a similar path in agricultural development over time and across countries. Pingali and Heisey (1996) categorized the technological transformation of cereal crop production system into three distinct phases:

i. the land-augmentation phase;

ii. the labour-substitution phase; and

iii. the knowledge-and management-intensity phase.

The basic assumption is that the transition from one phase to another is determined by growing factor scarcity, first for land, then for labour and finally for other factors of production, such as machinery and management skills.

The first phase is characterized by area expansion being the main source of output growth, as was seen during the pre-Green Revolution era of the 1950s and 1960s. However, as opportunities for area expansion decline over time, cropping intensity is increased, along with increasing use of water, fertilizers, pesticides and high yielding varieties.

This was indeed the case during the Green-Revolution period in the 1970s and early 1980s. Such intensive production results in an increased demand for labour and mechanization, as the production system moves from single-cropping to double-and triple-cropping with increased application of purchased inputs. Eventually, production reaches the point of diminishing marginal returns to further intensification, as was the case in the late 1980s, the post-Green Revolution phase.

Here, better technical knowledge and management skills are used to substitute for traditional inputs. Variety selection, fertilizer timing and placement, water management and pesticide application are some areas in which productivity has improved with reduction in unit cost of production.

The model just outlined is used in the following analysis as the basic framework to explain the changes in input use and in productivity between the 1960s and the 1990s. First, it is applied to various regions in the world, then to the developing and developed countries, and finally to the selected countries in Asia and the Pacific.

Comparisons of Agricultural Output and Input Use by Regions

Agricultural output along with usage rates for conventional inputs, including land, labour, tractor and fertilizer, in various regions are presented in Table 1. It can be seen that in 1994, Asia produced 47 percent of the world's agricultural output and had most of the agricultural (40 percent) and irrigated land (70 percent) and highest total fertilizer use (47 percent). Only in terms of total tractor use does it rank second, following Europe. In terms of input use per hectare, Asia ranks third for both fertilizer and tractors, following Europe and North America.

Notably, output in Asia tripled between 1961 and 1994. This three-fold increase can be attributed to the increase in input use (an 81 percent increase in irrigated land, a 13-fold increase in fertilizer usage, and a 27-fold increase in tractor usage).

However, there was only a slight increase (eight percent) in the amount of land used. In terms of factor productivity in Asia, it appears that fertilizer productivity declined by 80 percent from 1961 to 1994 while land productivity increased by 184 percent.

Because of the interaction between inputs used, these results should be interpreted with caution. Overall, the information presented in Table 1 suggests that although fertilizer usage in Asia is approaching the levels used in Europe and North America, land productivity can be improved further by increasing the degree of mechanization.

Table: Agricultural Output, Input Use and Productivity by Regions, 1961 and 1994

Input Use, 1961 and 1994					
	Africa	**Asia**	**Europe**	**Latin America**	**North America**
					1961
Net ag output (89-91 thousand million US$)	36.68	161.42	123.37	50.79	84.59
Land (million hectares)	155.12	436.21	151.37	102.27	225.71
Irrigated land (million hectares)	7.36	90.17	8.32	8.13	14.35
Percent of land irrigated	4.75	20.67	5.50	7.95	6.36
Tractors (million)	0.26	0.26	5.38	0.46	6.42
Fertilizer (million tonnes)	0.73	3.89	14.29	1.06	14.09
Fertilizer per hectare (kg)	4.68	8.92	94.43	10.33	62.45
Tractor per 1 000 ha (number)	1.65	0.60	35.53	4.53	28.45
					1994

Input Use, 1961 and 1994					
	Africa	Asia	Europe	Latin America	North America
Net ag output					
(89-91 thousand million US$)	78.81	496.74	186.14	122.42	162.68
Land (million hectares)	190.02	472.56	135.43	156.01	233.28
Irrigated land (million hectares)	12.20	163.17	16.77	17.65	22.11
Percent of land irrigated	6.42	34.53	12.38	11.32	9.48
Tractors (million)	0.59	7.87	12.35	1.67	6.36
Fertilizer (million tonnes)	3.47	61.08	21.98	9.23	32.86
Fertilizer per hectare (kg)	18.26	129.26	162.34	59.17	140.86
Tractor per 1 000 ha (number)	3.12	16.66	91.19	10.73	27.25

Factor Productivity, 1961 and 1994					
	Africa	Asia	Europe	Latin America	North America
				1961	
Net output/fertilizer (US$1 000/mt)	50.48	41.47	8.63	48.08	6.00
Net output/HA (US$/ha)	236.49	370.06	815.04	496.66	374.79
				1994	
Net output/fertilizer (US$1 000/mt)	22.71	8.13	8.47	13.26	4.95
Net output/HA (US$/ha)	414.72	1,051.18	1,374.48	784.70	697.37

Source: FAO (1997b).

Comparisons of Agricultural Output and Input Use, Developing Versus Developed Countries. In this section, comparisons were made for trends in land use, production and yield between developing and developed countries and across four commodities (rice, wheat, maize and other grains) for the period 1967-1994. The data were split into two subperiods, which coincided with the peak-green revolution period (1967-1982) and the post-green revolution period (1982-1994). The results are summarized in Table 2.

Table 2: Annual Growth Rates of Crop Area, Production and Yield 1967-1994 (percent)

	Area		Production		Yield		
	1967-1982	1982-1994	1967-1982	1982-1994	1967-1982	1982-1994	1993-2020P
			Wheat				
Developed	-0.12	-1.38	1.73	-0.03	1.87	1.35	1.06
Developing	1.45	0.42	5.39	2.94	3.88	2.52	1.30
World	0.48	-0.59	2.88	1.20	2.40	1.80	1.17
			Maize				

Contd...

	Area		Production		Yield		
Developed	0.64	-0.26	3.05	0.69	2.33	1.01	0.84
Developing	0.65	1.36*	3.46	3.66*	2.80	2.27	1.36
World	0.64	0.77*	3.20	1.93	2.52	1.16	1.03
			Paddy Rice				
Developed	-0.23	-0.28	-0.14	0.34*	0.09	0.61*	0.53
Developing	0.81	0.21	3.21	2.03	2.38	1.81	1.08
World	0.78	0.20	2.96	1.94	2.17	1.74	1.05
			Other Grains				
Developed	0.52	-1.63	1.32	-0.78	0.79	0.85*	0.78
Developing	-0.87	0.12	1.20	0.03	2.08	-0.09	1.24
World	-0.15	-0.79	1.28	-0.52	1.43	0.26	0.85

Source: Rosegrant and Ringler (1997).

p indicates projected, not observed, figures.

* indicates increases in growth rates between 1967-82 and 1982-94.

Several points can be drawn from Table 2. First, there are substantial variations in growth rates across all commodities, between developing and developed countries, and over time between the two subperiods. Secondly, the growth rates of cereal production and yield show a significant slowdown nearly across the board in the second subperiod. Thirdly, the crop area has been declining, again with only a few exceptions. Finally, the growth rates in yield between 1993 and 2020 are projected to be lower than what they were previously, with the exception of "other grains." Those exceptional cases are highlighted with asterisks in Table 2.

According to Rosegrant and Ringler (1997), the reduction in land area and production of wheat and other grains in developed countries was primarily policy-induced. It reflects the changes in price support programmes in North America and the Common Agricultural Policy in the European Union, as well as economic and political reforms in the formally centrally planned economies of Eastern Europe and the former Soviet Union. On the other hand, the slowdown of cereal productivity growth in developing countries, particularly in Asia, since the 1980s, was attributed to declining world prices and over-intensification of cereal production. Specifically, declining cereal prices had caused a shift of land out of cereals and into more profitable cropping alternatives, such as horticultural products. Furthermore, the intensity of land use in the late 1960s and 1970s, when the Green Revolution was in full swing, led to input usage beyond the optimal levels, reducing yield in the later period. From these observations, it seems apparent that land area available for cropping is unlikely to increase and may fall even further as more agricultural land is diverted to residential and industrial uses. A decrease in cropping area means that a greater burden will be placed on growth in crop yield to meet future cereal demand. Moreover, it appears that over-intensification may have led to resource degradation and hence a slowdown in yield growth. The implication is that, to maintain or increase yield in the future, more emphasis on sustainable agriculture is essential.

The Asian-Pacific Countries

The data presented here focus on seven Asian-Pacific countries and, where appropriate, use the United States as a benchmark. Moreover, comparisons are made based on agriculture as a whole rather than by commodity. The countries included are Australia, the United States, China, India, Indonesia, Japan, and South Korea.

Trends in input use, in terms of arable land and labour, and final agricultural output are presented in Table 3. The growth rates in labour use are variable. While there tended to be negative or little growth in labour input use in the United States, Japan, Australia and South Korea, there was slight to moderate growth in China, India and Indonesia. In terms of arable land use, the overall picture displayed little to negative growth. In addition, the negative growth was quite substantial in Indonesia and South Korea during 1987-94. The latter result could be attributable to fast industrialization in these two countries.

Table: Annual Average Growth Rates of Labour, Land Use and Agricultural Output (percent)

	USA	Japan	Australia	China	Indonesia	India	S. Korea
				Labour			
1961-75	-3.27	-3.66	-0.39	3.39	0.51	2.22	0.79
1975-87	-2.48	-2.53	0.22	0.59	1.60	1.50	-3.40
1987-94	0.86	-7.29	-1.10	0.98	2.08	2.55	-4.54
				Arable Land			
1961-75	0.23	-1.55	2.43	-0.39	0.00	0.36	0.10
1975-87	-0.03	-0.53	0.89	-0.38	1.36	0.11	-0.22
1987-94	0.00	-0.64	0.02	-0.23	-2.98	0.00	-1.17
				Final Agricultural Output			
1961-75	2.26	2.42	3.61	4.09	2.91	2.42	4.65
1975-87	1.00	0.86	0.93	4.99	4.42	2.58	3.10
1987-94	2.71	-0.42	1.73	5.15	3.38	4.64	2.29

Source: Rao and Lee (1997); FAO (1997b).

Agricultural output growth has remained positive from 1961 to 1994 (Table 3), with only one exception (Japan). Comparisons of growth rates are more variable, however.

For example, during 1975-87, the developed economies (United States, Japan and Australia) experienced a slowdown in production while production in less developed economies (China, India, South Korea and Indonesia) accelerated (Table 3).

The slowdown during the period 1975-1987 coincides with the growth deceleration in OECD countries during 1973-1987 in response to the oil shocks and resulting changes in macroeconomic policies (Maddison, 1989).

Maddison claims that Asian countries did not suffer as much from the oil price increases because of generally more flexible commodity and labour markets and less institutional rigidity that magnify external price shocks compared to OECD countries.

Another reason for strong growth in Asian countries was because of high levels of investment and rising educational levels during that period.

After the recession, during 1987-1994, Japan continued its slide to register a negative output growth (-0.42 percent), the United States showed a strong recovery and Australia recovered but only to half the rate it was before the slowdown.

In comparison, China and India had shown strong output growth throughout the observation period.

Also, it can be seen from Table 3 that, although during 1987-1994 the output growth rate was negative (-0.42 percent) in Japan, labour productivity grew by an impressive 7.41 percent, the highest among the countries listed.

The increase in productivity stemmed from the fact that the reduction in output was more than offset by the savings in labour use.

In China, output grew at a rate of 4.09 percent per year during 1961-1975 while labour productivity grew only marginally at a rate of 0.68 percent.

The differing rates imply that increases in output may have been due to greater use of labour rather than from productivity increases. In contrast, output growth in Japan comes mainly from productivity growth.

Table: Annual Average Growth Rates in Labour and Land Productivity (percent) (1987 Geary-Khamis Prices with Shadow Prices)

	USA	Japan	Australia	China	Indonesia	India	S. Korea
			Labour Productivity				
1961-75	5.72	6.32	4.02	0.68	2.39	0.20	3.83
1975-87	3.57	3.48	0.72	4.37	2.77	1.07	6.74
1987-94	1.84	7.41	2.87	4.13	1.28	2.04	7.16
			Land Productivity				
1961-75	2.03	4.03	1.16	4.51	2.91	2.05	4.54
1975-87	1.03	1.40	0.05	5.39	3.02	2.47	3.33
1987-94	2.71	0.22	1.71	5.39	6.56	4.65	3.51

Source: Rao and Lee (1997); FAO (1997b).

Table 4 shows the growth rates for partial factor productivity in terms of land and labour. It should be noted that both China and India showed strong growth in output between 1961 and 1994 (4.09 to 5.15 percent per annum for China and 2.42 to 4.64 percent for India, Table 3) but the results for productivity growth were somewhat different. In particular, China registered a dramatic increase in labour productivity from 0.68 percent per annum in 1961-1975 to 4.37 percent in 1975-1987 and 4.13 percent in 1987-1994; the comparable figures for India were 0.20, 1.07 and 2.04 percent. In general, output growth in both countries was due to the increased use of water and purchased inputs such as high yielding varieties and fertilizers (Wong, 1989).

South Korea and Japan both encountered a slowdown in output growth between 1961-1975 and 1987-1994, but labour productivity grew at impressive rates of 7.41 percent in Japan and 7.16 percent in South Korea during 1987-1994. Despite their high levels of labour productivity, Australia and the United States both encountered a slowdown during the period 1975-1987 in output and labour productivity growth. Although labour productivity picked up again during the period 1987-1994 for Australia, the United States continued its decline. Growth rates of land productivity, presented in Table 4, show that in general less developed countries experienced higher growth than developed countries. Japan from 1961 to 1975 is an exception. This could be the result of land intensification where input-intensive multiple cropping was a common practice to compensate for the scarcity of cultivated land. The national index of multiple cropping in China was 1.31 in 1952, 1.5 in 1978 and 1.58 in 1996 (Lin, 1998). In comparison, the index in India was 1.18 in 1970 and 1.24 in 1980 (Wong, 1989). These figures may explain some of the growth in land productivity in these two countries. Wong (1989) also claims that land reforms in China from collectivization to private ownership had a larger impact on land productivity than in India where land reform changed the land tenure system but not ownership. Despite strong growth in output and partial factor productivity of labour and land, total factor productivity growth has been found to be negative for China and India (Wong, 1989). These results indicate clearly that output growth was generated primarily from the expansion of inputs, rather than productivity increases.

Table 5 shows the relative labour productivity and yields in the selected Asian-Pacific countries using the 1961 figures for the United States as a benchmark. First, it can be seen that most countries have become more productive over time. Secondly, there are wide disparities among countries with three possible divisions. The first group includes the United States and Australia; the second, Japan and South Korea; and the third, China, Indonesia and India. For example, in 1961, agricultural labour in the United States was two-thirds as productive as that in Australia, nearly 15 to 25 times as productive as that in Japan and South Korea and more than 30 to 40 to times as productive as that in China, Indonesia and India. In 1994, the agricultural labour force in the United States was equally productive as that of Australia, 10 to 20 times as productive as Japanese and Korean agricultural labour and 60 to 80 times as productive as Chinese, Indonesian and Indian agricultural labour. The differences in mechanization and labour quality may explain the differences in the level of and changes in productivity in these countries.

Table: Indices of Agricultural Labour and Land Productivity in Selected Countries (USA 1961 = 100)

	USA	Japan	Australia	China	Indonesia	India	S. Korea
			Labour	**Productivity**			
1961	100.0	6.1	149.2	2.3	2.9	3.4	3.6
1975	217.9	14.3	258.9	2.5	4.0	3.5	6.2
1978	252.3	15.1	334.4	2.8	4.3	4.0	8.1
1987		21.6	282.1	4.2	5.5	3.9	13.5
1994	377.3	35.7	343.8	5.6	6.0	4.5	21.8
			Land	**Productivity**			
1961	100.0	415.9	51.9	06.1	106.0	63.6	227.0
1975	132.5	723.5	60.9	196.7	158.3	84.6	422.8
1978	138.1	751.5	74.4	218.4	174.5	93.3	525.6
1987	149.9	855.2	61.3	369.4	226.3	113.3	626.4
1994	180.8	868.4	69.0	533.3	352.8	155.7	797.3

Source: Rao and Lee (1997); FAO (1997b).

The comparative performance in terms of land productivity (or yield) of arable land shows a different picture (Table 5). In this case, Japan and South Korea have the highest yields, followed by China, then by Indonesia, the United States, India and Australia. These results indicate that countries with limited land resources tend to farm their lands more intensively and hence have higher output per unit of arable land. They also reflect differences in climate and water availability. In summary, it appears that there are substantial variations in input use and productivity among countries and over time. The only trend that is common to the countries examined is perhaps the negative growth in arable land use. Secondly, there are substantial variations in the level of productivity and the rate of productivity growth among developing countries (China, India, and Indonesia), newly industrialized countries (Japan and South Korea) and developed countries (Australia and the United States).

This suggests that there is ample room for productivity improvements in the less developed countries. Meanwhile, the gaps in productivity, as presented in Table 5, are closing. Furthermore, the fact that growth in output and various measures of productivity can sometimes move in opposite directions confirms the important distinctions between output growth and productivity growth, and between total and partial factor productivity.

Finally, it is apparent that the slowing or negative growth in global crop area will increasingly place the burden of meeting future cereal demand on productivity improvements. Productivity improvement can come either from using existing inputs more efficiently (moving closer to the production frontier) or from technological change (shifting the production frontier upward), or a combination of both. It has been shown that, with existing technology, efficiency is primarily influenced by human capital, such as farmers' education and experience, access to credit and extension services (Coelli and Battese, 1996). Technological change depends, on the other hand, on investments in agricultural research and extension (Alston, Norton and Pardey, 1995; Antle and Capalbo, 1988; Pray and Evenson, 1991). In the next section, sources of productivity growth are discussed based on a survey of existing literature.

Factors Affecting Productivity Growth

In explaining productivity growth, economists originally limited themselves to the role of conventional inputs such as land, labour, physical capital, water and chemical inputs. However, the failure to explain productivity growth adequately led them to examine the role of human capital and public goods, such as education, agricultural research and extension and publicly provided infrastructure (Griliches, 1963; Mankiw, Romer and Weil, 1992; Nelson, 1964 and 1981; Solow, 1957). Public policies that have a strong link to agricultural productivity such as policy reforms were also examined.

The rationale for considering research is the belief that investments in research result in increases in the stock of knowledge, which, in turn, either facilitate the use of existing knowledge or generate new technology.

Technological advances, whether resulting from changes in input quality or how inputs are combined, lead to productivity gains. Education, training and extension also increase productivity by increasing people's knowledge and skill base, which are essential for technology adoption and efficient use of inputs. Public infrastructure, on the other hand, increases productivity by facilitating the exchange of goods and services.

Technological Change

Technological change is recognized by many as one of the most important sources of productivity growth (Antle and Capalbo, 1988). It refers to the changes in the production process that come about from the application of innovation and newly acquired scientific knowledge and technical and management skills.

Technological change increases agricultural productivity either by shifting the production frontier upward so that more measured output can be produced with the same amount of inputs or by moving closer to the production frontier so that the same amount of output can be produced with a smaller amount of inputs. Better organizational and management skills not only improve input-output combinations but enable producers to respond more quickly to changing market circumstances.

While generation of new technology or knowledge comes from investments in research and development, adoption of technology involves investments by the potential users in both physical and human capital. Therefore, adoption of technology depends principally on their applicability and expected returns of the innovation. However, there may be a long lag between development, adoption and productivity gains. Chavas and Cox (1992) found the lag to be up to 15 years between making an investment in research and having an effect on productivity. However, after taking effect, the benefits from an innovation may persist for thirty years or more. The lag between generation of new technology and its widespread adoption by farmers has important policy implications.

First, the adverse effects of reduced public funding to agricultural research and extension on productivity may be under-estimated if the lagged effects are not accounted for. Secondly, the complementarity between research and extension should be taken into account.

The former helps the development of new technology, while the latter helps speed up the rate of diffusion and adoption of new technology. Extension can be done more effectively by identifying factors that contribute to technology adoption. As an example, innovators in a farming community can be identified and targeted for extension services.

Since better-educated farmers are found to be more likely to adopt new technology, human capital is a pre-condition for technology adoption and hence productivity growth. Further, if adoption of new technology requires additional investments, lack of access to credit and additional inputs may prevent or slow down technology adoption. Finally, because potential users of new technology often differ in the agronomic-ecological conditions in which they operate, new technology may require adaptive research before it can be transferred successfully to different locations. These impediments to technology adoption mean careful planning and provision of necessary infrastructure are essential to capture the full benefits of new technology.

Agricultural Research and Extension

Many researchers have explored the roles of research and extension in promoting agricultural growth. Rosegrant and Evenson (1992) found that in South Asia, public research accounted for 30 percent of the output growth, and extension for about 25 percent, with corresponding rates of return being 63 percent and 52 percent, respectively. Pray and Evenson's (1991) survey of Asia found the rates of return to national research investment ranged from 19 to 218 percent, returns to national extension investment from 15 to 215 percent, and returns to international research investment from 68 to 108 percent.

Evenson and McKinsey (1991) found that public investment in research accounted for over half of the output growth in India and extension contributed about one-third. The calculated internal rates of return were 218 percent for public research and 177 percent for extension. However, they found that little output growth was attributable to infrastructure. Jamison and Lau (1982) also found that physical capital had little impact on production or profits, as compared to farmer's education and extension services.

Fan (1996) found that public research expenditures accounted for about 20 percent of total production growth in Chinese agriculture during the period 1965 to 1994. The annual rates of return to agricultural research investment in China ranged from 44 percent to 83 percent. Fan (1996) concluded that the rapid growth in agricultural output in China during the 1980s and 1990s was the result of public investments in R&D as well as the institutional and market reforms that began in 1979. He concluded that increases in agricultural research were justifiable; not only did they stimulate additional output growth, but the rate of return to agricultural research was much higher than commercial interest rates.

Despite the high rates of returns from public research investments, agricultural research intensity (ARI), measured as a percentage of Chinese agricultural GDP, was found to have declined from 0.56 percent for the period 1958-1965 to 0.43, 0.44, 0.39 and 0.40 percent, respectively, for 1966-1976, 1977-1985, 1986-1990 and 1991-1993 (Fan, 1996). Lin (1998) reported that, as part of the overall market reform, the Chinese Government had reduced its fiscal appropriation for agricultural research, shifting funding from institutional supports to competitive grants and cost recovery. As such, it can be expected that an increasing proportion of research activities will move from the public to the private domain. Other studies on output growth have also shown a high payoff from agricultural research and extension. The results indicate that the rate of return on research, in most cases, ranged from 15 to 50 percent for both developed and developing countries, but some estimates were as high as 218 percent. The wide disparity among the estimates raises questions regarding the sensitivity of these estimates to the commodity of interest and the use of different time periods and methodologies.

Estimates for Asian countries appear to be higher and show a much wider variation than those of studies in the United States. This could be due to the diverse nature of Asian agriculture, which differs from country to country in economic, social and agronomic-ecological conditions. Because of inconsistency in the data and methodology used, it is not possible to make direct comparisons across countries or over time. Nevertheless, the general conclusion that R&D yields relatively high returns seems indisputable.

Human Capital

Human capital refers to knowledge, experience and skills possessed by people involved in the production process. It is influenced directly by education, training and extension. Its importance lies in the fact that it has a significant impact on the adoption and the utilization of technology, which in turn, affect the allocation of resources and productivity. A well-trained and well-educated labour force is said to be in a better position to assess changing conditions and make necessary adjustments. This ability is becoming ever more important in an increasingly deregulated and global economy where changes in the commodity markets are frequent and quick responses are required.

The concept of investment in human capital covers not only investments in formal schooling and post-school and on-the-job training, but also investment in the form of improved health and family care. Social capital, on the other hand, refers to one's ability to utilize social networks and institutions. Social status, education and the range of social institutions available can influence one's social capital. Social capital is important in that it affects access to physical capital, land title, credit and cooperatives, all of which have implications for resource allocation and, hence, productivity. Women appear to suffer most severely from having limited access to human and social capital in some developing countries, although a larger proportion of women than men are engaged in agriculture (Quisumbing *et al.*, 1995). For example, women account for 70 to 80 percent of household food production in sub-Saharan Africa, 65 percent in Asia, and 45 percent in Latin America and the Caribbean.

As such, it has been suggested that empowerment of women and gender equality are important factors for raising productivity and promoting food security in developing countries. Jahnke, Kirschke and Lagemann (1987) also found that low adoption of high yield varieties (HYV) in Africa is attributable to lack of appropriate technology development and few extension services directed to women. These findings suggest that acknowledging the critical role of rural women in Asia and providing them with greater access to resources and human capital are crucial in promoting sustainable agriculture and food security.

7

Sustainable Agriculture and Environment in Forestry Sector

The uniqueness of microorganisms and their often unpredictable nature and biosynthetic capabilities, given a specific set of environmental and cultural conditions, has made them likely candidates for solving particularly difficult problems in the life sciences and other fields as well. The various ways in which microorganisms have been used over the past 50 years to advance medical technology, human and animal health, food processing, food safety and quality, genetic engineering, environmental protection, agricultural biotechnology, and more effective treatment of agricultural and municipal wastes provide a most impressive record of achievement. Many of these technological advances would not have been possible using straightforward chemical and physical engineering methods, or if they were, they would not have been practically or economically feasible.

Nevertheless, while microbial technologies have been applied to various agricultural and environmental problems with considerable success in recent years, they have not been widely accepted by the scientific community because it is often difficult to consistently reproduce their beneficial effects. Microorganisms are effective only when they are presented with suitable and optimum conditions for metabolizing their substrates Including available water, oxygen (depending on whether the microorganisms are obligate aerobes or facultative anaerobes), pH and temperature of their environment. Meanwhile, the various types of microbial cultures and inoculants available in the market today have rapidly increased because of these new technologies. Significant achievements are being made in systems where technical guidance is coordinated with the marketing of microbial products. Since microorganisms are useful in eliminating problems associated with the use of chemical fertilizers and pesticides, they are now widely applied in nature farming and organic agriculture.

Environmental pollution, caused by excessive soil erosion and the associated transport of sediment, chemical fertilizers and pesticides to surface and groundwater, and improper treatment of human and animal wastes has caused serious environmental and social problems throughout the world. Often engineers have attempted to solve these problems using established chemical and physical methods. However, they have usually found that such problems cannot be solved without using microbial methods and technologies in coordination with agricultural production.

For many years, soil microbiologists and microbial ecologists have tended to differentiate soil microorganisms as beneficial or harmful according to their functions and how they affect soil quality, plant growth and yield, and plant health. Beneficial microorganisms are those that can fix atmospheric nitrogen, decompose organic wastes and residues, detoxify pesticides, suppress plant diseases and soil-borne pathogens, enhance nutrient cycling, and produce bioactive compounds such as vitamins, hormones and enzymes that stimulate plant growth. Harmful microorganisms are those that can induce plant diseases, stimulate soil-borne pathogens, immobilize nutrients, and produce toxic and putrescent substances that adversely affect plant growth and health.

A more specific classification of beneficial microorganisms has been suggested by Higa (1991; 1994; 1995) which he refer to as "Effective Microorganisms" or EM. This report presents some new perspectives on the role and application of beneficial microorganism, including EM, as microbial inoculants for shifting the soil microbiological equilibrium in ways that can improve soil quality, enhance crop production and protection, conserve natural resources, and ultimately create a more sustainable agriculture and environment The report also discusses strategies on how beneficial microorganisms, including EM. can be more effective after inoculation into soils.

The Concept of Effective Microorganisms: their Role and Application

The concept of effective microorganisms (EM) was developed by Professor Teruo Higa, University of the Ryukyus, Okinawa, Japan.

EM consists of mixed cultures of beneficial an naturally-occurring microorganisms that can be applied as inoculants to increase the microbial diversity of soils and plant. Research has shown that the inoculation of EM cultures to the soil/plant ecosystem can improve soil quality, soil health, and the growth, yield, and quality of crops. EM contains selected species of microorganisms including predominant populations of lactic acid bacteria and yeasts and smaller numbers of photosynthetic bacteria, actinomycetes and other types of organisms. All of these are mutually compatible with one another and can coexist in liquid culture.

EM is not a substitute for other management practices. It is, however, an added dimension for optimizing our best soil and crop management practices such as crop rotations, use of organic amendments, conservation tillage, crop residue recycling, and biocontrol of pests. If used properly, EM can significantly enhance the beneficial effects of these practices.

Throughout the discussion which follows, we will use the term "beneficial microorganisms" In a general way to designate a large group of often unknown or ill-defined microorganisms that interact favourably in soils and with plants to render beneficial effects which are sometimes difficult to predict. We use the term "effective microorganisms" or EM to denote specific mixed cultures of known, beneficial microorganisms that are being used effectively as microbial inoculants.

Utilization of Beneficial Microorganisms in Agriculture

What Constitutes an Ideal Agricultural System?

Conceptual design is important in developing new technologies for utilizing beneficial and effective microorganisms for a more sustainable agriculture and environment. The basis of a conceptual design is imply to first conceive an ideal or model and then to devise a strategy and method for achieving the reality. However it is necessary to carefully coordinate the materials, the environment, and the technologies constituting the method. Moreover one should adopt a philosophical attitude in applying microbial technologies to agricultural production and conservation systems.

There are many opinions on what an ideal agricultural system is. Many would agree that such an idealized system should produce food on a long-term sustainable basis. Many would also insist that it should maintain and improve human health, be economically and spiritually beneficial to both producers and consumers, actively preserve and protect the environment, be self-contained and regenerative, and produce enough food for an increasing world population.

Efficient Utilization and Recycling of Energy

Agricultural production begins with the process of photosynthesis by green plants which requires solar energy, water, and carbon dioxide. It occurs through the plants ability to utilize solar energy in "fixing" atmospheric carbon into carbohydrates. The energy obtained is used for further biosynthesis in the plant, including essential amino acids and proteins. The materials used for agricultural production are abundantly available with little initial cost. However, when it is observed as an economic activity, the fixation of carbon by photosynthesis has an extremely low efficiency mainly because of the low utilization rate of solar energy by green plants. Therefore, an integrated approach is needed to increase the level of solar energy utilization by plants so that greater amounts of atmospheric carbon can be converted into useful substrates.

Although the potential utilization rate of solar energy by plants has been estimated theoretically at between 10 and 20%, the actual utilization rate is less than 1%. Even the utilization rate of C4 plants, such as sugar cane whose photosynthetic efficiency is very high, barely exceeds 6 or 7% during the maximum growth period. The utilization rate is normally less than 3% even for optimum crop yields. Past studies have shown that photosynthetic efficiency of the chloroplasts of host crop plants cannot be increased much further; this means that their biomass production has reached a maximum level. Therefore, the best opportunity for increasing biomass production is to somehow utilize the visible light, which chloroplasts cannot presently use, and the infrared radiation; together, these comprise about 80% of the total solar energy. Also, we must explore ways of recycling organic energy contained in plant and animal residues through direct utilization of organic molecules by plants.

Thus, it is difficult to exceed the existing limits of crop production unless the efficiency of utilizing solar energy is increased, and the energy contained in existing organic molecules (amino acids, peptides and carbohydrates) is utilized either directly or indirectly by the plant. This approach could help to solve the problems of environmental pollution and degradation caused by the misuse and excessive application of chemical fertilizers and pesticides to soils. Therefore, new technologies that can enhance the economic-viability of farming systems with little or no use of chemical fertilizers and pesticides are urgently needed and should be a high priority of agricultural research both now and in the immediate future.

Preservation of Natural Resources and the Environment

The excessive erosion of topsoil from farmland caused by intensive tillage and row-crop production has caused extensive soil degradation and also contributed to the pollution of both surface and groundwater. Organic wastes from animal production, agricultural and marine processing industries, and municipal wastes (i.e., sewage and garbage), have become major sources of environmental pollution in both developed and developing countries. Furthermore, the production of methane from paddy fields and ruminant animals and of carbon dioxide from the burning of fossil fuels, land clearing and organic matter decomposition have been linked to global warming as "greenhouse gases".

Chemical-based, conventional systems of agricultural production have created many sources of pollution that, either directly or indirectly, can contribute to degradation of the environment and destruction of our natural resource base. This situation would change significantly if these pollutants could be utilized in agricultural production as sources of energy. Therefore, it is necessary that future agricultural technologies be compatible with the global ecosystem and with solutions to such problems in areas different from those of conventional agricultural technologies. An area that appears to hold the greatest promise for technological advances in crop production, crop protection, and natural resource conservation is that of beneficial and effective microorganisms applied as soil, plant and environmental inoculants (Higa, 1995).

Beneficial and Effective Microorganisms for a Sustainable Agriculture

Agriculture in a broad sense, is not an enterprise which leaves everything to nature without intervention. Rather it is a human activity in which the farmer attempts to integrate certain agroecological factors and production inputs for optimum crop and livestock production. Thus, it is reasonable to assume that farmers should be interested in ways and means of controlling beneficial soil microorganisms as an important component of the agricultural environment. Nevertheless, this idea has often been rejected by naturalists and proponents of nature farming and organic agriculture. They argue that beneficial soil microorganisms will increase naturally when organic amendments are applied to soils as carbon, energy and nutrient sources.

This indeed may be true where an abundance of organic materials are readily available for recycling which often occurs in small-scale farming. However, in most cases, soil microorganisms, beneficial or harmful, have often been controlled advantageously when crops in various agroecological zones are grown and cultivated in proper sequence (i.e., crop rotations) and without the use of pesticides. This would explain why scientists have long been interested in the use of beneficial microorganisms as soil and plant inoculants to shift the microbiological equilibrium in a way that enhances soil quality and the yield and quality of crops.

Most would agree that a basic rule of agriculture is to ensure that specific crops are grown according to their agroclimatic and agroecological requirements. However, in many cases the agricultural economy is based on market forces that demand a stable supply of food, and thus, it becomes necessary to use farmland to its full productive potential throughout the year.

The purpose of crop breeding is to improve crop production, crop protection, and crop quality. Improved crop cultivars along with improved cultural and management practices have made it possible to grow a wide variety of agricultural and horticultural crops in areas where it once would not have been culturally or economically feasible. The cultivation of these crops in such diverse environments has contributed significantly to a stable food supply in many countries. However, it is somewhat ironic that new crop cultures are almost never selected with consideration of their nutritional quality or bioavailability after ingestion (Hornick, 1992).

As will be discussed later, crop growth and development are closely related to the nature of the soil microflora, especially those in close proximity to plant roots, i.e., the rhizosphere. Thus, it will be difficult to overcome the limitations of conventional agricultural technologies without controlling soil microorganisms. This particular tenet is further reinforced because the evolution of most forms of life on earth and their environments are sustained by microorganisms.

Most biological activities are influenced by the state of these invisible, minuscule units of life. Therefore, to significantly increase food production, it is essential to develop crop cultivars with improved genetic capabilities (i.e., greater yield potential, disease resistance, and nutritional quality) and with a higher level of environmental competitiveness, particularly under stress conditions (i.e., low rainfall, high temperatures, nutrient deficiencies, and agressive weed growth).

To enhance the concept of controlling and utilizing beneficial microorganisms for crop production and protection, one must harmoniously integrate the essential components for plant growth and yield including light (intensity, photoperiodicity and quality), carbon dioxide, water, nutrients (organic-inorganic) soil type, and the soil microflora. Because of these vital interrelationships, it is possible to envision a new technology and a more energy-efficient system of biological production.

Low agricultural production efficiency is closely related to a poor coordination of energy conversion which, in turn, is influenced by crop physiological factors, the environment, and other biological factors including soil microorganisms. The soil and rhizosphere microflora can accelerate the growth of plants and enhance their resistance to disease and harmful insects by producing bioactive substances. These microorganisms maintain the growth environment of plants, and may have secondary effects on crop quality. A wide range of results are possible depending on their predominance and activities at any one time. Nevertheless, there is a growing consensus that it is possible to attain maximum economic crop yields of high quality, at higher net returns, without the application of chemical fertilizers and pesticides. Until recently, this was not thought to be a very likely possibility using conventional agricultural methods. However, it is important to recognize that the best soil and crop management practices to achieve a more sustainable agriculture will also enhance the growth, numbers and activities of beneficial soil microorganisms that, in turn, can improve the growth, yield and quality of crops.

Controlling the Soil Microflora: Principles and Strategies

Principles of Natural Ecosystems and the Application of Beneficial and Effective Microorganisms

The misuse and excessive use of chemical fertilizers and pesticides have often adversely affected the environment and created many a) food safety and quality and b) human and animal health problems.

Consequently, there has been a growing interest in nature farming and organic agriculture by consumers and environmentalists as possible alternatives to chemical-based, conventional agriculture.

Agricultural systems which conform to the principles of natural ecosystems are now receiving a great deal of attention in both developed and developing countries. A number of books and journals have recently been published which deal with many aspects of natural farming systems. New concepts such as alternative agriculture, sustainable agriculture, soil quality, integrated pest management, integrated nutrient management and even beneficial microorganisms are being explored by the agricultural research establishment. Although these concepts and associated methodologies hold considerable promise, they also have limitations. For example, the main limitation in using microbial inoculants is the problem of reproducibility and lack of consistent results.

Unfortunately certain microbial cultures have been promoted by their suppliers as being effective for controlling a wide range of soil-borne plant diseases when in fact they were effective only on specific pathogens under very specific conditions. Some suppliers have suggested that their particular microbial inoculant is akin to a pesticide that would suppress the general soil microbial population while increasing the population of a specific beneficial microorganism. Nevertheless, most of the claims for these single-culture microbial inoculants are greatly exaggerated and have not proven to be effective under field conditions. One might speculate that if all of the microbial cultures and inoculants that are available as marketed products were used some degree of success might be achieved because of the increased diversity of the soil microflora and stability that is associated with mixed cultures. While this, of course, is a hypothetical example, the fact remains that there is a greater likelihood of controlling the soil microflora by introducing mixed, compatible cultures rather than single pure cultures.

Even so, the use of mixed cultures in this approach has been criticized because it is difficult to demonstrate conclusively which microorganisms are responsible for the observed effects, how the introduced microorganisms interact with the indigenous species, and how these new associations affect the soil/plant environment. Thus, the use of mixed cultures of beneficial microorganisms as soil inoculants to enhance the growth, health, yield, and quality of crops has not gained widespread acceptance by the agricultural research establishment because conclusive scientific proof is often lacking. The use of mixed cultures of beneficial microorganisms as soil inoculants is based on the principles of natural ecosystems which are sustained by their constituents; that is, by the quality and quantity of their inhabitants and specific ecological parameters, i.e., the greater the diversity and number of the inhabitants, the higher the order of their interaction and the more stable the ecosystem. The mixed culture approach is simply an effort to apply these principles to natural systems such as agricultural soils, and to shift the microbiological equilibrium in favour of increased plant growth, production and protection.

It is important to recognize that soils can vary tremendously as to their types and numbers of microorganisms. These can be both beneficial and harmful to plants and often the predominance of either one depends on the cultural and management practices that are applied. It should also be emphasized that most fertile and productive soils have a high content of organic matter and, generally, have large, populations of highly diverse microorganisms (i.e., both species and genetic diversity). Such soils will also usually have a wide ratio of beneficial to harmful microorganisms.

Controlling the Soil Microflora for Optimum Crop Production and Protection

The idea of controlling and manipulating the soil microflora through the use of inoculants organic amendments and cultural and management practices to create a more favorable soil microbiological environment for optimum crop production and protection is not new.

For almost a century, microbiologists have known that organic wastes and residues, including animal manures, crop residues, green manures, municipal wastes (both raw and composted), contain their own indigenous populations of microorganisms often with broad physiological capabilities.

It is also known that when such organic wastes and residues are applied to soils many of these introduced microorganisms can function as biocontrol agents by controlling or suppressing soil-borne plant pathogens through their competitive and antagonistic activities.

While this has been the theoretical basis for controlling the soil microflora, in actual practice the results have been unpredictable and inconsistent, and the role of specific microorganisms has not been well-defined.

For, many years microbiologists have tried to culture beneficial microorganisms for use as soil inoculants to overcome the harmful effects of phytopathogenic organisms, including bacteria, fungi and nematodes. Such attempts have usually involved single applications of pure cultures of microorganisms which have been largely unsuccessful for several reasons. First, it is necessary to thoroughly understand the individual growth and survival characteristics of each particular beneficial microorganism, including their nutritional and environmental requirements. Second, we must understand their ecological relationships and interactions with other microorganisms, including their ability to coexist in mixed cultures and after application to soils.

There are other problems and constraints that have been major obstacles to controlling the microflora of agricultural soils. First and foremost is the large number of types of microorganisms that are present at any one time, their wide range of physiological capabilities, and the dramatic fluctuations in their populations that can result from man's cultural and management practices applied to a particular farming system. The diversity of the total soil microflora depends on the nature of the soil environment and those factors which affect the growth and activity of each individual organism including temperature, light, aeration, nutrients, organic matter, pH and water. While there are many microorganisms that respond positively to these factors, or a combination thereof, there are many that do not. Microbiologists have actually studied relatively few of the microorganisms that exist in most agricultural soil, mainly because we don't know how to culture them; i.e., we know very little about their growth, nutritional, and ecological requirements.

The "diversity" and "population" factors associated with the soil microflora have discouraged scientists from conducting research to develop control strategies. Many believe that, even when beneficial microorganisms are cultured and inoculated into soils, their number is relatively small compared with the indigenous soil inhabitants, and they would likely be rapidly overwhelmed by the established soil microflora. Consequently, many would argue that even if the application of beneficial microorganisms is successful under limited conditions (e.g., in the laboratory) it would be virtually impossible to achieve the same success under actual field conditions. Such thinking still exists today, and serves as a principle constraint to the concept of controlling the soil microflora.

It is noteworthy that most of the microorganisms encountered in any particular soil are harmless to plants with only a relatively few that function as plant pathogens or potential pathogens. Harmful microorganisms become dominant if conditions develop that are favorable to their growth, activity and reproduction. Under such conditions, soil-borne pathogens (e.g., fungal pathogens) can rapidly increase their populations with devastating effects on the crop. If these conditions change, the pathogen population declines just as rapidly to its original state. Conventional farming systems that tend toward the consecutive planting of the same crop (i.e., monoculture) necessitate the heavy use of chemical fertilizers and pesticides.

This, in turn, generally increases the probability that harmful, disease-producing, plant pathogenic microorganisms will become more dominant in agricultural soils. Chemical-based conventional farming methods are not unlike symptomatic therapy. Examples of this are applying fertilizers when crops show symptoms of nutrient-deficiencies, and applying pesticides whenever crops are attacked by insects and diseases. In efforts to control the soil microflora some scientists feel that the introduction of beneficial microorganisms should follow a symptomatic approach. However, we do not agree. The actual soil conditions that prevail at any point in time may be most unfavourable to the growth and establishment of laboratory-cultured, beneficial microorganisms. To facilitate their establishment, it may require that the farmer make certain changes in his cultural and management practices to induce conditions that will (a) allow the growth and survival of the inoculated microorganisms and (b) suppress the growth and activity of the indigenous plant pathogenic microorganisms.

An example of the importance of controlling the soil microflora and how certain cultural and management practices can facilitate such control is useful here. Vegetable cultivars are often selected on their ability to grow and produce over a wide range of temperatures. Under cool, temperate conditions there are generally few pest and disease problems. However, with the onset of hot weather, there is a concomitant increase in the incidence of diseases and insects making it rather difficult to obtain acceptable yields without applying pesticides. With higher temperatures, the total soil microbial population increases as does certain plant pathogens such as Fusarium, which is one of the main putrefactive, fungal pathogens in soil. The incidence and destructive activity of this pathogen can be greatly minimized by adopting reduced tillage methods and by shading techniques to keep the soil cool during hot weather. Another approach is to inoculate the soil with beneficial, antagonistic, antibiotic-producing microorganisms such as actinomycetes and certain fungi.

Application of Beneficial and Effective Microorganisms: A New Dimension

Many microbiologists believe that the total number of soil microorganisms can be increased by applying organic amendments to the soil. This is generally true because most soil microorganisms are heterotrophic, i.e., they require complex organic molecules of carbon and nitrogen for metabolism and biosynthesis.

Whether the regular addition of organic wastes and residues will greatly increase the number of beneficial soil microorganisms in a short period of time is questionable.

However, we do know that heavy applications of organic materials, such as seaweed, fish meal, and chitin from crushed crab shells, not only helps to balance the micronutrient content of a soil but also increases the population of beneficial antibiotic-producing actinomycetes. This changes the soil to a disease-suppressive condition within a relatively short period.

The probability that a particular beneficial microorganism will become predominant, even with organic farming or nature farming methods, will depend on the ecosystem and environmental conditions. It can take several hundred years for various species of higher and lower plants to interact and develop into a definable and stable ecosystem. Even if the population of a specific microorganism is increased through cultural and management practices, whether it will be beneficial to plants is another question.

Thus, the likelihood of a beneficial, plant-associated microorganism becoming predominant under conservation-based farming systems is virtually impossible to predict. Moreover, it is very unlikely that the population of useful anaerobic microorganisms, which usually comprise only a small part of the soil microflora, would increase significantly even under natural farming conditions.

This information then emphasizes the need to develop methods for isolating and selecting different microorganisms for their beneficial effects on soils and plants. The ultimate goal is to select microorganisms that are physiologically and ecologically compatible with each other and that can be introduced as mixed cultures into soil where their beneficial effects can be realized.

Application of Beneficial and Effective Microorganisms: Fundamental Considerations

Microorganisms are utilized in agriculture for various purposes; as important components of organic amendments and composts, as legume inoculants for biological nitrogen fixation as a means of suppressing insects and plant diseases to Improve crop quality and yields, and for reduction of labour. All of these are closely related to each other. An important consideration in the application of beneficial microorganisms to soils is the enhancement of their synergistic effects. This is difficult to accomplish if these microorganisms are applied to achieve symptomatic therapy, as in the case of chemical fertilizers and pesticides.

If cultures of beneficial microorganisms are to be effective after inoculation into soil, it is important that their initial populations be at a certain critical threshold level. This helps to ensure that the amount of bioactive substances produced by them will be sufficient to achieve the desired positive effects on crop production and/or crop protection.

If these conditions are not met, the introduced microorganisms, no matter how useful they are, will have little if any effect. At present, there are no chemical tests that can predict the probability of a particular soil-inoculated microorganism to achieve a desired result. The most reliable approach is to inoculate the beneficial microorganism into soil as part of a mixed culture, and at a sufficiently high inoculum density to maximize the probability of its adaptation to environmental and ecological conditions.

The application of beneficial microorganisms to soil can help to define the structure and establishment of natural ecosystems. The greater the diversity of the cultivated plants that are grown and the more chemically complex the biomass, the greater the diversity of the soil microflora as to their types, numbers and activities. The application of a wide range of different organic amendments to soils can also help to ensure a greater microbial diversity. For example, combinations of various crop residues, animal manures, green manures, and municipal wastes applied periodically to soil will provide a higher level of microbial diversity than when only one of these materials is applied. The reason for this is that each of these organic materials has its own unique indigenous microflora which can greatly affect the resident soil microflora after they are applied, at least for a limited period.

Classification of Soils Based on Their Microbiological Properties

Most soils are classified on the basis of their chemical and physical properties; little has been done to classify soils according to their physicochemical and microbiological properties. The reason for this is that a soil's chemical and physical properties are more readily defined and measured than their microbiological properties. Improved soil quality is usually characterized by increased infiltration; aeration, aggregation and organic matter content and by decreased bulk density, compaction, erosion and crusting.

While these are important indicators of potential soil productivity, we must give more attention to soil biological properties because of their important relationship (though poorly understood) to crop production, plant and animal health, environmental quality, and food safety and quality.

Research is needed to identify and quantify reliable and predictable biological/ecological indicators of soil quality. Possible indicators might include total species diversity or genetic diversity of beneficial soil microorganisms as well as insects and animals.

The basic concept here is not to classify soils for the study of microorganisms but for farmers to be able to control the soil microflora so that biologically-mediated processes can improve the growth, yield, and quality of crops as well as the tilth, fertility, and productivity of soils. The ultimate objective is to reduce the need for chemical fertilizers and pesticides.

Functions of Microorganisms: Putrefaction, Fermentation, and Synthesis

Soil microorganisms can be classified into decomposer and synthetic microorganisms. The decomposer microorganisms are subdivided into groups that perform oxidative and fermentative decomposition. The fermentative group is further divided into useful fermentation (simply called fermentation) and harmful fermentation (called putrefaction). The synthetic microorganisms can be sub-divided into groups having the physiological abilities to fix atmospheric nitrogen into amino acids and/or carbon dioxide into simple organic molecules through photosynthesis. Adapted from Higa is a simplified flow chart of organic matter transformations by soil microorganisms that can lead to the development of disease-inducing, disease-suppressive, zymogenic, or synthetic soils.

Fermentation is an anaerobic process by which facultative microorganisms (e.g., yeasts) transform complex organic molecules (e.g., carbohydrates) into simple organic compounds that often can be absorbed directly by plants.

Fermentation yields a relatively small amount of energy compared with aerobic decomposition of the same substrate by the same group of microorganisms. Aerobic decomposition results in complete oxidation of a substrate and the release of large amounts of energy, gas, and heat with carbon dioxide and water as the end products. Putrefaction is the process by which facultative heterotrophic microorganisms decompose proteins anaerobically, yielding malodorous incompletely oxidized, metabolites (e.g., ammonia, mercaptans and indole) that are often toxic to plants and animals.

The term "synthesis" as used here refers to the biosynthetic capacity of certain microorganisms to derive metabolic energy by "fixing" atmospheric nitrogen and/or carbon dioxide. In this context we refer to these as "synthetic" microorganisms, and if they should become a predominant part of the soil microflora, then the soil would be termed a "synthetic" soil. Nitrogen-fixing microorganisms are highly diverse, ranging from "free-living" autotrophic bacteria of the genus Azotobacter to symbiotic, heterotrophic bacteria of the genus Rhizobium, and blue-green algae (now mainly classified as blue-green bacteria), all of which function aerobically. Photosynthetic microorganisms fix atmospheric carbon dioxide in a manner similar to that of green plants. They are also highly diverse, ranging from blue-green algae to green algae that perform complete photosynthesis aerobically to photosynthetic bacteria which perform incomplete photosynthesis anaerobically.

Relationships Between Putrefaction, Fermentation, and Synthesis

The processes of putrefaction, fermentation, and synthesis proceed simultaneously according to the appropriate types and numbers of microorganisms that are present in the soil. The impact on soil quality attributes and related soil properties is determined by the dominant process.

The production of organic substances by microorganisms results from the intake of positive ions, while decomposition serves to release these positive ions. Hydrogen ions play a pivotal role in these processes. A problem occurs when hydrogen ions do not recombine with oxygen to form water but are utilized to produce methane, hydrogen sulfide, ammonia, mercaptans and other highly reduced putrefactive substances most of which are toxic to plants and produce malodors. If a soil is able to absorb the excess hydrogen ions during periods of soil anaerobiosis and if synthetic microorganisms such as photosynthetic bacteria are present, they will utilize these putrefactive substances and produce useful substrates from them which helps to maintain a healthy and productive soil.

The photosynthetic bacteria, which perform incomplete photosynthesis anaerobically, are highly desirable, beneficial soil microorganisms because they are able to detoxify soils by transforming reduced, putrefactive substances such as hydrogen sulfide into useful substrates. This helps to ensure efficient utilization of organic matter and to improve soil fertility. Photosynthesis involves the photo-catalyzed splitting of water which yields molecular oxygen as a by-product. Thus, these microorganisms help to provide a vital source of oxygen to plant roots.

Reduced compounds such as methane and hydrogen sulfide are often produced when organic materials are decomposed under anaerobic conditions. These compounds are toxic and can greatly suppress the activities of nitrogen-fixing microorganisms. However, if synthetic microorganisms, such as photosynthetic bacteria that utilize reduced substances, are present in the soil, oxygen deficiencies are not likely to occur. Thus, nitrogen-fixing microorganisms, coexisting in the soil with photosynthetic bacteria, can function effectively in fixing atmospheric nitrogen even under anaerobic conditions.

Photosynthetic bacteria not only perform photosynthesis but can also fix-nitrogen. Moreover, it has been shown that, when they coexist, in soil with species of Azotobacter, their ability to fix nitrogen is enhanced. This then is an example of a synthetic soil. It also suggests that by recognizing the role, function, and mutual compatibility of these two bacteria and utilizing them effectively to their full potential, soils can be induced to a greater synthetic capacity. Perhaps the most effective synthetic soil system results from the enhancement of zymogenic and synthetic microorganisms; this allows fermentation to become dominant over putrefaction and useful synthetic processes to proceed.

Classification of Soils Based on the Functions of Microorganisms

As discussed earlier, soils can be characterized according to their indigenous microflora which perform putrefactive, fermentative, synthetic and zymogenic reactions and processes. In most soils, these three functions are going on simultaneously with the rate and extent of each determined by the types and numbers of associated microorganisms that are actively involved at any one time.

A simple diagram showing a classification of soils based on the activities and functions of their predominant microorganisms is presente.

Disease-Inducing Soils. In this type of soil, plant pathogenic microorganisms such as Fusarium fungi can comprise 5 to 20 percent of the total microflora if fresh organic matter with a high nitrogen content is applied to such a soil, incompletely oxidized products can arise that are malodorous and toxic to growing plants. Such soils tend to cause frequent infestations of disease organisms, and harmful insects. Thus, the application of fresh organic matter to these soils is often harmful to crops. Probably more than 90 percent of the agricultural land devoted to crop production worldwide can be classified as having disease-inducing soil. Such soils generally have poor physical properties, and large amounts of energy are lost as "greenhouse" gases, particularly in the case of rice fields. Plant nutrients are also subject to immobilization into unavailable forms.

Disease-Suppressive Soils. The microflora of disease-suppressive soils is usually dominated by antagonistic microorganisms that produce copious amounts of antibiotics. These include fungi of the genera Penicillium, Trichoderma, and Aspergillus, and actinomycetes of the genus Streptomyces. The antibiotics they produce can have biostatic and biocidal effects on soil-borne plant pathogens, including Fusarium which would have an incidence in these soils of less than 5 percent. Crops planted in these soils are rarely affected by diseases or insect pests. Even if fresh organic matter with a high nitrogen content is applied, the production of putrescent substances is very low and the soil has a pleasant earthy odor after the organic matter is decomposed. These soils generally have excellent physical properties; for example, they readily, form water-stable aggregates and they are well-aerated, and have a high permeability to both air and water. Crop yields in the disease-suppressive soils are often slightly lower than those in synthetic soils. Highly acceptable crop yields are obtained whenever a soil has a predominance of both disease-suppressive and synthetic microorganisms.

Zymogenic Soils. These soils are dominated by a microflora that can perform useful kinds of fermentations, i.e., the breakdown of complex organic molecules into simple organic substances and inorganic materials. The organisms can be either obligate or facultative anaerobes. Such fermentation-producing microorganisms often comprise the microflora of various organic materials, i.e., crop residues, animal manures, green manures and municipal wastes including composts. After these amendments are applied to the soil, their number: and fermentative activities can increase dramatically and overwhelm the indigenous soil microflora for an indefinite period. While these microorganisms remain predominant, the soil can be classified as a zymogenic soil which is generally characterized by a) pleasant, fermentative odours especially after tillage, b) favorable soil physical properties (e.g., Increased aggregate stability, permeability, aeration and decreased resistance to tillage c) large amounts of inorganic nutrients, amino acids, carbohydrates, vitamins and other bioactive substances which can directly or indirectly enhance the growth, yield and quality of crops, d) low occupancy of Fusarium fungi which is usually less than 5 percent, and e) low production of greenhouse gases (e.g., methane, ammonia, and carbon dioxide) from croplands, even where flooded rice is grown.

Synthetic Soils. These soils contain significant populations of microorganisms which are able to fix atmospheric nitrogen and carbon dioxide into complex molecules such as amino acids, proteins and carbohydrates.

Such microorganisms include photosynthetic bacteria which perform incomplete photosynthesis anaerobically, certain Phycomycetes (fungi that resemble algae), and both green algae and blue—green algae which function aerobically. All of these are photosynthetic organisms that fix atmospheric nitrogen. If the water content of these soils is stable, their fertility can be largely maintained by regular additions of only small amounts of organic materials. These soils have a low Fusarium occupancy and they are often of the disease-suppressive type. The production of gases from fields where synthetic soils are present is minimal, even for flooded rice.

This is a somewhat simplistic classification of soils based on the functions of their predominant types of microorganisms, and whether they are potentially beneficial or harmful to the growth and yield of crops. While these different types of soils are described here in a rather idealized manner, the fact is that in nature they are not always clearly defined because they often tend to have some of the same characteristics. Nevertheless, research has shown that a disease-inducing soil can be transformed into disease-suppressing, zymogenic and synthetic soils by inoculating the problem soil with mixed cultures of effective microorganisms. Thus it is somewhat obvious that the most desirable agricultural soil for optimum growth, production, protection, and quality of crops would be the composite soil, i.e., a soil that is highly zymogenic and synthetic, and has an established disease-suppressive capacity. This then is the principle reason for seeking ways and means of controlling the microflora of agricultural soils.

Summary and Conclusions

Controlling the soil microflora to enhance the predominance of beneficial and effective microorganisms can help to improve and maintain the soil chemical and physical properties. The proper and regular addition of organic amendments are often an important part of any strategy to exercise such control.

Previous efforts to significantly change the indigenous microflora of a soil by introducing single cultures of extrinsic microorganisms have largely been unsuccessful. Even when a beneficial microorganism is isolated from a soil, cultured in the laboratory, and reinoculated into the same soil at a very high population, it is immediately subject to competitive and antagonistic effects from the indigenous soil microflora and its numbers soon decline.

Thus, the probability of shifting the "microbiological equilibrium" of a soil and controlling it to favour the growth, yield and health of crops is much greater if mixed cultures of beneficial and effective microorganisms are introduced that are physiologically and ecologically compatible with one another. When these mixed cultures become established their individual beneficial effects are often magnified in a synergistic manner.

Actually, a disease-suppressive microflora can be developed rather easily by selecting and culturing certain types of gram-positive bacteria that produce antibiotics and have a wide range of specific functions and capabilities; these organisms include facultative anaerobes, obligate aerobes, acidophilic and alkalophilic microbes. These microorganisms can be grown to high populations in a medium consisting of rice bran, oil cake and fish meal and then applied to soil along with well-cured compost that also has a large stable population of beneficial microorganisms, especially facultative anaerobic bacteria. A soil can be readily transformed into a zymogenic/synthetic soil with disease-suppressive potential if mixed cultures of effective microorganisms with the ability to transmit these properties are applied to that soil. The desired effects from applying cultured beneficial and effective microorganisms to soils can be somewhat variable, at least initially. In some soils, a single application (i.e., inoculation) may be enough to produce the expected results, while for other soils even repeated applications may appear to be ineffective.

The reason for this is that in some soils it takes longer for the introduced microorganisms to adapt to a new set of ecological and environmental conditions and to become well-established as a stable, effective and predominant part of the indigenous soil microflora.

The important consideration here is the careful selection of a mixed culture of compatible, effective microorganisms properly cultured and provided with acceptable organic substrates. Assuming that repeated applications are made at regular intervals during the first cropping season, there is a very high probability that the desired results will be achieved.

There are no meaningful or reliable tests for monitoring the establishment of mixed cultures of beneficial and effective microorganisms after application to a soil. The desired effects appear only after they are established and become dominant, and remain stable and active in the soil.

The inoculum densities of the mixed cultures and the frequency of application serve only as guidelines to enhance the probability of early establishment. Repeated applications, especially during the first cropping season, can markedly facilitate early establishment of the introduced effective microorganisms.

Once the "new" microflora is established and stabilized, the desired effects will continue indefinitely and no further applications are necessary unless organic amendments cease to be applied, or the soil is subjected to severe drought or flooding.

Finally, it is far more likely that the microflora of a soil can be controlled through the application of mixed cultures of selected beneficial and effective microorganisms than by the use of single or pure cultures. If the microorganisms comprising the mixed culture can coexist and are physiologically compatible and mutually complementary, and if the initial inoculum density is sufficiently high, there is a high probability that these microorganisms will become established in the soil and will be effective as an associative group, whereby such positive interactions would continue. If so, then it is also highly, probable that they will exercise considerable control over the indigenous soil microflora which, in due course, would likely be transformed into or replaced by a "new" soil microflora.

Integrating Seasonal Climate Prediction and Agricultural Models

There is a strong history of integrated climate–crop models to estimate the impacts of climate change scenarios, and adaptive breeding and land use strategies. Advances in climate prediction at a seasonal lead-time have stimulated substantial interest in forecasting crop yields as a means of improving farm management and policy-level interventions in a manner that reduces risk and enhances livelihoods and food security, particularly in the marginal, rainfed, tropical cropping regions that the green revolution largely bypassed. Farmers experience climate change not as a shifting mean but as climatic variations. Climatic variability at all of its time-scales is, therefore, a current challenge to development. Furthermore, developing flexible, proactive strategies for managing year-to-year climate variations within farming communities and the institutions that interface with them, using advance climate information, is arguably the most concrete step that the agricultural enterprise can take to build resilience to long-term changes in the global climate system. These two realizations appear to be bringing the historically distinct global change adaptation and agricultural development agendas closer together, particularly in Africa.

Several research groups are seeking to advance methodology to integrate crop simulation models with dynamic climate forecast models for application at a range of spatial scales. A range of promising approaches has emerged (Hansen *et al.* submitted*b*) that may have relevance to management of climate variability and application to climate change projections. In this paper, we seek to put such work in context by discussing how integrated climate–crop models operating at a seasonal lead-time might contribute to efforts to address climate-related constraints to agricultural development, with some emphasis on rainfed agriculture in Africa. Section 2 summarizes the role of climatic risk and potential role of seasonal climate prediction in agriculture. Section 3 relates integrated climate–crop modelling to constraints that have been identified to effective use of seasonal forecasts within agriculture. The three subsequent sections propose three areas in which integrated modelling might contribute to more effective use of seasonal climate forecasts for the benefit of smallholder agriculture in the developing countries.

Climate Risk and Agricultural Development

Climate exerts a profound influence on the lives of poor rural populations who depend on agriculture for livelihood and sustenance, who are unprotected against climate-related diseases, who lack secure access to water and food, and who are vulnerable to hydrometeorological hazard. The year-to-year variability of rainfall is a significant constraint to the sustainability of rainfed farming systems in poorer countries of the tropics. This argument is developed more fully in Hansen *et al.* submitted*a*. Climatic extremes, such as drought and flooding, have a direct and often persistent impact on farmers' assets and livelihoods. Perhaps, more important (Elbers *et al.* 2003), the uncertainty associated with climate variability creates a moving target for management, and is a disincentive to sustainable resource management, and to the intensification and adoption of innovation that is necessary for secure livelihoods and long-term rural prosperity....Variability by itself is not necessarily welfare-decreasing if it is anticipated and acted upon.

Surprise, however, has adverse consequences since the optimal *ex post* and *ex ante* choices rarely coincide Rainfall variability tends to be the dominant source of livelihood risk in smallholder rainfed agriculture, particularly in dryer environments. In the face of climatic uncertainty, risk-averse farmers employ conservative strategies, including avoidance of improved technology, under-use of fertilizers and shifting from productive to non-productive liquid assets. In sub-Sahara Africa, where soil nutrient depletion is now recognized as a root cause of declining per capita food production and a critical constraint to sustainable livelihoods, the apparent effect of climatic uncertainty on investment in soils is a particular concern. These *ex ante* strategies, designed to buffer against climatic extremes, substantially reduce average income and marginal productivity of assets, and do so disproportionately for the relatively poor.

Despite the tremendous concern about climate change, climate variability has not received as much attention as other development issues, in part because it has been considered part of the environmental baseline that is not amenable to intervention. Climate variations in the months ahead no longer need to be accepted as a total unknown.

The chaotic nature of the atmosphere restricts deterministic prediction of weather events to lead-times of several days. However, year-to-year variations in the atmosphere are influenced by interactions between the atmosphere and the more slowly varying ocean and land surfaces, such as those associated with the El Nino-Southern Oscillation (ENSO) in the tropical Pacific. Improvements in our understanding of interactions between the atmosphere and its underlying surfaces, advances in modelling the global climate system, and substantial investment in monitoring the tropical oceans now provide a degree of predictability of climate fluctuations at a seasonal (i.e. several months) lead-time in many parts of the world. Goddard *et al.* (2001) provide a useful, recent review of the scientific basis and methodology of seasonal climate prediction. Where the necessary conditions are in place or can be put into place, seasonal climate prediction offers an under-exploited opportunity to manage climate variability: to respond proactively to adverse conditions and exploit favourable conditions.

Integrated Climate–crop Modelling and the Climate Risk Problem

The potential for seasonal forecasts to reduce the adverse impacts of climate variability and enhance rural livelihoods has motivated a number of pilot studies targeting smallholder farmers in developing countries. Several pilot studies with African farmers have demonstrated a high level of interest and have identified a range of promising livelihood management responses. They have also identified several obstacles to using seasonal forecasts effectively to reduce the adverse impacts of climate variability and enhance rural livelihoods. I highlight three for which integrated climate–crop modelling has a potential role. The first is a mismatch between farmers' needs and the scale, relevance and transparency of available forecasts.

The second is the cost, risk and learning time farmers face during adaptation and adoption. Third, propose that failure to obtain adequate institutional, financial and political support is also a key constraint. Farm-scale studies have identified several resource constraints that limit responses to advance climate information, but that might be amenable to intervention if additional institutional actors (e.g. suppliers of production inputs and credit) are engaged. Scaling up beyond plot studies will require substantial investment on the part of national agricultural and meteorological services. Providing comprehensive institutional support, including provision of relevant climate information, appropriate technical guidance and intervention in financial services and production inputs (Hansen *et al.* submitted*a*) will require the involvement of a broader range of institutions that have typically been involved in pilot studies.

Several research groups who are working on applications of seasonal climate forecasts have invested heavily in the use of crop simulation models. Although, the use of weather data from analogue years associated with, e.g. ENSO phases, has been the standard approach, interest in coupling crop simulation with dynamic climate forecast models has increased substantially in recent years. A mismatch between the scale of dynamic climate models and crop simulation models complicates the task.

Crop growth depends more on the distribution of weather within a season than on the season averages that forecasters typically provide. Crop simulation models, generally, assume the small-scale of a homogenous plot, and simulate dynamic interactions between weather, management and crop growth and development on a daily time-step. On the other hand, the spatial resolution of the current generation of general circulation models (GCMs) is on the order of 10 000 km^2. Although, GCMs simulate the atmosphere on a sub-daily time-step, the spatial averaging within grid cells distorts day-to-day variability, with potentially serious consequences for simulated crop responses. Fortunately, recent and ongoing research has given rise to several promising approaches to translating seasonal climate forecasts into probabilistic predictions of crop response (reviewed in Hansen *et al.* submitted*b*).

It is clear that applications of seasonal climate forecasts to manage risk are concerned with impacts on, e.g. crops and not climate *per se*. Yet, the most critical opportunities and challenges have more to do with institutional support than technology. If that is the case, then what contribution can agricultural models integrated with seasonal climate prediction models make? I propose three promising contributions: (i) translating seasonal climate forecasts into useful information; (ii) *ex ante* assessment of potential benefits and (iii) fostering and guiding application.

Translating Climate Forecasts into Useful Information

Hammer (2000) argued that seasonal climate forecasts have no intrinsic value. First, value comes from improved decisions not information. Second, it is not climatic means but impacts within the system being managed that are relevant to decisions. If climate forecasts are to have value, they must modify decision makers' expectations of the production or economic outcomes that relate to goals, within the system being managed (Luseno *et al.* 2003). Crop models integrated with seasonal climate forecasts provide a means of translating forecasts of seasonal climate anomalies into forecasts of production impacts. The nature of the decision determines the appropriate spatial scale and lead-time of crop forecast information.

Farm Level Applications

Producers of annual crops must routinely make a range of critical production and livelihood decisions prior to planting that interact with climate, but whose outcome is not realized until harvest several months later. Where predictability is sufficient, seasonal forecasts may provide probabilistic information about crop yields with sufficient lead-time to influence pre-planting decisions.

A mismatch between the content, scale, format and lead-time of available operational climate forecasts, and the information that farmers need for such decisions, is the most often cited constraint to effective use. It is also the constraint that is most within the control of climate information providers.

Although, farmers' information requirements are somewhat context-specific, a few generalizations emerge from experience in a range of contexts. Farmers need information that: (i) can be interpreted at a field scale; (ii) includes information about timing beyond three-month climatic means; (iii) is explicit about accuracy; and (iv) is expressed in terms of impacts and management implications within the agricultural systems that they manage.

In contrast, operational climate forecasts are typically expressed as three month climatic means averaged over large areas. When accuracy is communicated, it is typically expressed as shifts in the probability of experiencing outcomes within each climatological tercile, with no information about how the probabilities are derived.

The widespread awareness associated with the 1997 El Nino event stimulated a great deal of interest and debate about the potential value of climate prediction to poor farmers in developing countries. Some expressed doubts about the feasibility of meeting farmers' information requirements.

Is predictability of climate and crop response at the farm scale sufficient to be useful? If not, will responding to forecasts that could be wrong expose farmers to unacceptable risk? Barrett (1998) expressed concern that the level of predictability of crop yields at a farm scale is likely to be inadequate for use by risk-averse smallholder farmers.

First, fine-scale spatial variability of rainfall implies that seasonal rainfall predictability is limited to aggregate spatial scales. Second, because crop yield is not a simple function of seasonal total rainfall, the accumulation of errors going from seasonal climatic predictors (e.g. sea surface temperatures), to local seasonal means, to crop response implies that predictions of impacts such as crop response will be less accurate than predictions of climatic means. Barrett called for greater investment in methods for translating climate forecasts into forecasts of relevant farm-level impacts.

Averaging in either time or space tends to increase prediction skill by reducing the random 'noise' component of variability in weather. As Gong *et al.* (2003) demonstrate, the skill of GCM-based seasonal climate forecasts tends to increase with increasing spatial aggregation relative to a single GCM grid cell in several parts of the world.

For the state of Ceará, in northeast Brazil, they also showed that downscaling to individual single stations reduces prediction skill, but that a substantial proportion of the predictability remains. Although limited experience seems to support the generalization that forecasts that are skilful at an aggregate scale show only moderate decline at individual points, there remains a need to evaluate the influence of scale of aggregation on prediction skill systematically in a range of locations.

The argument that predictability of agricultural impacts is necessarily less than that of climatic means overlooks two considerations.

First, information (e.g. antecedent rainfall, stored soil moisture) beyond seasonal climatic anomalies can contribute to predictability of crop response. Second, predicting relevant impacts directly from climatic predictors instead of from predicted climatic means reduces accumulation of errors, and potentially incorporates information about rainfall distribution and other relevant meteorological variables that are embedded in climatic predictors, but lost when converting them into seasonal rainfall totals.

In a paper that arguably stimulated much of the interest in using seasonal forecasts to benefit farmers in Africa, Cane *et al.* (1994) showed that ENSO-related sea surface temperatures in the Pacific were correlated more strongly with maize yields (1970–1993) than with rainfall averaged across Zimbabwe. Dilley (1997) found that maize yields in Oaxaca, Mexico, were correlated more strongly with the Southern Oscillation Index than with local rainfall. K. P. C. Rao, the APSRU model to simulate maize yields for the short rains season at Katumani, Kenya, as a function of observed weather and GCM-based monthly hindcasts disaggregated to daily values using a stochastic weather generator.

Correlations between predictions and observations were higher for simulated yields than for seasonal rainfall totals, although the analysis did not account for crop model error. By integrating GCM-based seasonal rainfall forecasts into an operational regional wheat forecasting system in Queensland, Australia, Hansen *et al.* (2004) showed that reported district yields were more predictable than seasonal rainfall totals.

They attributed this in part to the influence of observed antecedent rainfall and resulting soil moisture storage on subsequent yields. I anticipate that methodological advances and empirical evidence will address concerns about the predictability of crop response to climate fluctuations. Yet, given the spatial heterogeneity, multiplicity of yield-reducing factors and paucity of data in rainfed farming systems of much of Africa, realistic, site-specific forecasts of crop yields are likely to be elusive for some time.

Early Warning, Market and Policy Applications

Advance information about crop production is relevant to decision-makers operating at a range of scales from field to national or regional, whose decisions impact the welfare of farmers. Institutions responsible for responding to drought or food crises and traders and investors in commodity markets, are interested in anticipating production at relatively large spatial scales.

Because of the adverse impacts of variability on commodity prices, the use of advance information about production to manage markets for staple foods to stabilize prices is appealing from the standpoint of food security. Although, experience with using climate information to intervene in markets is still limited, economy-wide modelling in Mozambique suggests considerable potential aggregate benefits of market applications of climate forecasts.

In food-insecure regions of Africa, early warning systems based on remote sensing of rainfall and vegetation, and monitoring of local commodity markets, provide early indication of likely food shortfalls. When a crisis requires external food aid, it can take several months to verify shortfalls, mobilize donors, transport food to the region and distribute it to the affected population. As a result, assistance often reaches the affected people after they have already suffered adverse health effects, divested productive resources or migrated away from their farms.

Food aid organizations in Africa are increasingly interested in seasonal climate forecasts because of their potential to increase lead-time. For example, in August 2004, stakeholders concerned with food security early warning and response in East Africa initiated a periodic food security outlook forum, associated with the Greater Horn of Africa Climate Outlook Forum, with the goal of projecting likely food security impacts associated with the climate outlook to inform response.

In contrast to many climate-sensitive decisions at the farm level, market and food security early warning applications can often use forecast information at various points within the growing season.

The combination of increased spatial scale (Gong *et al.* 2003) and usefulness of forecasts at shorter lead-times increases the predictability potentially available for such applications. On the other hand, institutional constraints to responding to food crises are likely to require a higher threshold of forecast certainty than individual livelihood decisions.

Operational crop forecasting systems exist in many regions to support market and policy applications and food security early warning systems. These are often based on monitoring weather and crop conditions during the growing season.

In some instances, they incorporate regionally calibrated crop models to simulate growth or water stress up to the current point within the growing season, then sample weather for the remainder of the season from past years to estimate yield forecast uncertainty. Skilful seasonal forecasts provide additional information that can increase the accuracy of within-season production forecasts based on monitoring and simulation alone, particularly early in the growing season. Integrating probabilistic forecasts of food production from integrated climate–crop models into models of household vulnerability has potential to improve aid assessments and targeting at a relatively long lead-time.

Ex Ante Assessment of Benefit

Ex ante impact assessment seeks to assess the potential outcomes of an innovation in advance of its adoption, while *ex post* assessment seeks to assess actual outcomes following adoption. Because the use of seasonal climate forecasts within agriculture is a relatively new innovation, *ex ante* methods are, in most cases, the only way to estimate their benefits.

Pilot studies have compiled some evidence of use of forecasts for farm decisions, but have, generally, not tried to quantify the resulting production or livelihood benefits. Even after farmers learn to use forecast information routinely, *ex post* assessment of benefits would require multiple years due to the stochastic nature of climate variability and forecast responses—longer than typical funding cycles allow.

Furthermore, there are few, if any, regions in the developing world where rural communities have had access to operational climate information tailored to their needs for sufficient time to allow *ex post* assessment of use and benefits.

Ex ante impact assessment serves two related roles in applied agricultural research: providing evidence of potential benefit of proposed interventions to support mobilization of resources, and providing insight to target interventions where the positive impact is likely to be greatest (Thornton submitted). Resource mobilization has to do with allocation of scarce resources between a particular intervention or institution, and others that compete. Targeting relates to an institution or program's allocation of scarce resources among its own competing activities.

Confidence and Credibility

Increasing competition for dwindling resources for applied research has driven the international agricultural research community to become proactive about both quantifying the benefits of past activities (i.e. *ex post* impact assessment), and providing evidence of the likely benefits of proposed interventions (i.e. *ex ante* assessment). Likewise, donors increasingly require *ex ante* evidence of benefit before they will invest heavily in new agricultural innovations. The application of seasonal climate prediction is a new innovation that potentially competes for scarce resources with technologies such as crop genetic improvements that have relatively long records of *ex post* evidence of impact. On the positive side, it shares with integrated natural resource management (INRM) a relatively short history, heavy demands on management and context-specific implementation that prevent robust *ex post* impact evaluation (Barrett 2003). Yet, the broad suite of interventions that fall under INRM are generally well funded within the international agricultural research community.

Climate applications face the additional challenge of engaging climate and agricultural institutions that often do not have a history of substantive cooperation. Although, there are good theoretical reasons to expect that farmers and other agricultural stakeholders should benefit, there are still few well documented *ex post* demonstrations of adoption and benefit in developing countries, relative to many other agricultural development interventions. Some rather fundamental expressions of doubt about the potential utility to farmers in developing countries appear in the literature, citing examples of lack of use and barriers to use of available seasonal forecasts by smallholder farmers.

Hansen (2004) summarizes some of these concerns in a set of questions: is there enough predictability at the farm scale to be useful? Do smallholder farmers have the capacity to respond to climate forecasts? Will climate forecasts that could be wrong expose farmers to unacceptable risk? Can smallholder farmers with limited education understand probabilistic climate forecasts? Is the application of seasonal climate forecasts inherently biased against the poor? With the exception of farmers' ability to understand probabilistic forecasts, well designed, data-driven *ex ante* assessments can help answer these concerns for particular contexts. Credible evidence of acceptability and benefit to farmers may contribute to mobilizing funds and influencing the agendas of institutional partners. On the other hand, much of the support for forecast applications for agriculture has come from sources that do not traditionally fund agricultural research and development, and that have not yet demanded the level of evidence that traditional agricultural donors typically require.

Targeting Interventions

The potential to benefit from seasonal climate forecasts lies in the intersection of predictability of relevant components of climate variability, vulnerability (more correctly, sensitivity) to the adverse impacts of climate variability and ability to modify climate-sensitive decisions (Hansen 2002). *Ex ante* impact assessment, using integrated climate–crop–economic modelling, can provide a quantitative basis for selecting priorities for allocating scarce resources among regions, farming systems and interventions. Examples of the use of *ex ante* modelling to target regions or populations seem to be absent from the climate applications literature. There are, however, a few examples of analysis that are relevant to setting priorities for forecast system development and policy.

8

Forestry Resources

Land is among the most important resources used by U.S. agriculture—and its chief finite resource. Decisions on land use are made daily by operators of the Nation's 2.1 million farms, contingent on market conditions, public policies, and farm/ranch characteristics. This section examines land use, farm real estate values, and the relationship between land ownership, farm size, and other characteristics. Among the findings:

- Land continues to shift between agriculture and other uses. Cropland has declined over recent decades, but losses do not threaten the Nation's capacity to produce food and fibre.
- Real estate is a major asset for farm households. Values have been rising steadily since 1987, but the inflation-adjusted (real) value of U.S. farm real estate is still below its 1982 peak.
- Small family farms account for most land owned by farms.

Land Use

Land-use changes can affect the environment and the sustainability of production. Because impacts on the environment—including erosion, water quality, and wildlife habitat—are typically not reflected in private profit calculations, land-use choices that are optimal for an individual may not be optimal for society.

This difference suggests the possibility of public policies that more closely align land-use decisions with social objectives. The allocation of a fixed land base among competing uses is determined by the relative returns to the different uses, which vary according to land quality and location. A landowner seeking to maximize profits will allocate a land parcel to the use that yields the highest expected economic return, after the costs of conversion.

As relative returns change along with market conditions, technological advancements, or government policies, land-use patterns tend to adjust accordingly.

Land-use change is dynamic. With the exception of urban land, changes occur to and from major land uses. For example, 44 million acres left the cropland and pasture category from 1992 to 1997 while 21 million acres shifted into the category, resulting in a net loss of 23 million acres (USDA/NRCS, 2000).

Major Land Uses in the United States

Major Land Uses

This series contains acreage estimates of major uses by region and State, coinciding with each census of agriculture from 1945 through 2002. Because Alaska and Hawaii have very little crop area, we focus on the contiguous 48 States.

The total land area of the 48 contiguous States is approximately 1.9 billion acres, with an additional 365 million acres in Alaska and a little over 4 million acres in Hawai.

Grassland pasture and range, the largest use of land, accounted for 584 million acres (31 percent) of the 48 States in 2002. This compares with 636 million acres in the mid-1960s. Due to improvements in the forage quality and productivity of grazing lands, less pasture and range is needed to sustain grazing herds. The inventory of domestic animals, particularly sheep, has also been declining in recent years, further reducing pasture/range demand.

Forest-use land, the second largest major use, declined from about 32 percent of total land in 1945 to about 30 percent in 2002. A broader category, all land with forest cover, comprised 33 percent of the land base in 2002.

While forest-use land increased 1 percent between 1997 and 2002, it declined from 612 million acres in 1964 to 559 million acres in 2002. Much forest-covered land is in "special uses" that prohibit forestry uses such as timber production. Forested land in these special uses increased from 23 million acres in 1945 to about 98 million acres in 2002.

***Table:** Major uses of land, United States, 2002*[1]

Land use	48 States	US	48 States	US
	Million acres		% of total	
Cropland[2]	441	442	23.3	19.5
Grassland pasture and range	584	587	30.8	25.9
Forest-use land[3]	559	651	29.5	28.8
Special uses[4]	153	297	8.1	13.1
Urban	59	60	3.1	2.6
Miscellaneous other land	97	228	5.1	10.1
Total land area[5]	1,894	2,264	100.0	100.0

[1]See Major Land Uses for estimates of major uses by region and State, coinciding with each census of agriculture, from 1945 through 2002.

[2]All land in the crop rotation (used for crops, used for pasture, idle cropland). Includes about 34 million acres idled under the Conservation Reserve Program.

[3]Total forest land as classified by the U.S. Forest Service minus an estimated 98 million acres of forested land used for parks, wildlife areas, and other special uses.

[4]Rural transportation areas, land used primarily for recreation and wildlife purposes, various public installations and facilities, farmsteads, and farm roads/lanes. Excludes urban land in contrast to Major Land Uses, Aggregate Data.

[5]Distributions by major use may not add to totals due to rounding.

Sources: USDA/ERS based primarily on reports and records of the Census Bureau and Federal, State, and local land management and conservation agencies.

Cropland comprises the third largest use of land, covering 23 percent of the contiguous States in 2002. Since 1945, cropland ranged from a high of 478 million acres in 1949 to a low of 441 million acres in 2002. Total cropland has trended downward since the late 1960s, and decreased by 13 million acres (3 percent) from 1997 to 2002.

The total cropland base includes cropland used for crops, cropland used for pasture, and cropland idled. These components vary more than total cropland. Since 1945, the amount of cropland used for crops has ranged fromas much as 383 million acres in 1949 and 1982 to a minimum of 331 million acres in 1987. Total acreage used for crops exhibited two major cycles between 1945 and 1987, with cropland moving from idle to crop use and back again. Cropland used for crops increased from 331 to 349 million acres over 1987-97, and then declined to 340 million acres in 2002, about 5 percent below the average acreage for 1910-97. Since 1945, cropland used for pasture varied from 47 million acres in 1945 to 88 million acres in 1969.

Special uses include rural transportation; rural parks and wildlife; defence and industrial uses; and farmstead, farm roads/lanes, and other onfarm uses. These special uses increased from 85 million acres (4 percent of the land area of the contiguous 48 States) in 1945 to 153 million acres (8 percent) in 2002. Land in transportation uses (highways and roads, railroads, and airports in rural areas) increased by 4 million acres (17 percent) between 1945 and 1982. Transportation uses declined by about 0.5 million acres from 1982 to 1992 due to the abandonment of railroad facilities and rural roads, and the classification of some transportation uses as urban areas.

Land used for recreation and wildlife areas (Federal and State parks, wilderness areas, and wildlife refuges) expanded 344 percent from 1945 to 2002 (an increase of 78 million acres). The increase came mostly from conversion of Federal lands, previously in forest and grassland pasture and range. Land in defence and industrial uses declined by 10 million acres (40 percent) from 1945 to 2002. Farmsteads, farm roads, and other farm uses declined by 4 million acres (29 percent) between 1945 and 1997.

This decline reflects trends toward fewer farms and larger, more consolidated farms, as well as an increasing tendency for farm households to live off the farm. In response to expanding U.S. population, land in urban uses—including homes, schools, office buildings, shopping sites, and other commercial/industrial uses—increased from 15 million acres in 1945 to 25 million acres in 1960, 47 million acres in 1980, and 59 million acres in 2002. While the U.S. population nearly doubled, the amount of land urbanized quadrupled. However, urban uses still comprise only 3 percent of the total land area of the contiguous States.

Miscellaneous other land uses decreased from 1945 to 1964, and have since trended upward, showing a 54-percent increase from 1964 to 2002 reflecting improved data and reclassification of grazing and forest lands. These uses include marshes and open swamps not included in other major land uses, bare rock areas, deserts, some rural residential areas, and other uses not inventoried. Wetlands are defined by soil and hydrological characteristics and may occur on land in many different uses.

Regional Changes in Land Use

While land in every use occurs in all 10 regions of the contiguous States, some uses are more concentrated in some regions than in others.

Regions with the largest cropland acreage are the Northern Plains, Corn Belt, and Southern Plains. Grassland pasture and range is concentrated in the Mountain and Southern Plains regions. Acreage in forest use, special and miscellaneous other uses is highest in the Mountain region.

The Northeast, Appalachian, Southeast, Delta States, and Lake States regions lost cropland between 1945 and 2002. The largest increases occurred in the Northern Plains and Mountain regions, with smaller increases in the Corn Belt, Southern Plains, and Pacific regions. Western increases may have resulted in part from federally subsidized irrigation water.

Nine of the 10 regions lost grassland pasture and range between 1945 and 2002. While grassland pasture and range increased 11 million acres (10 percent) in the Southern Plains, the Northeast region lost about 70 percent of its grassland pasture and range, and the Appalachian and Lake States regions lost more than 50 percent.

The Northeast and Appalachian regions saw the reforestation of grassland, loss of some grassland to urbanization, and concentration of the dairy industry. Decreases in the Corn Belt, Northern Plains, and Mountain regions were likely associated with the conversion of some grassland pasture and range to cropland.

Cropland Use and Federal Programs

While total cropland acreage has varied up and down and generally declined since 1969, greater shifts have occurred between cropland used for crops and cropland idled, mostly because of Federal programs. Cropland used for pasture has exhibited less variation in acreage than cropland idled.

Most cropland used for crops is harvested, but typically 2-3 percent experiences crop failure and 5-10 percent is cultivated summer fallow. In 2002, farmers harvested one or more crops on an estimated 307 million acres of cropland, down 4 percent from 1997. About 8 million acres of the total harvested were double-cropped. When double-cropped land is counted twice, total acres harvested rise to 315 million acres.

Table: Major uses of cropland, 48 contiguous States, 1992-2004

Cropland	1992	1997	2002	2004[1]
		Million acres		
Cropland used for crops[2]	337	349	340	336
Cropland harvested[3]	305	321	307	312
Crop failure	8	7	17	9
Cultivated summer fallow	24	21	16	15
Cropland idled by all Federal programs[2]	55	33	34	35
Annual programs	19	0	0	0
Conservation Reserve Program	35	33	34	35
Total, specified uses[2,4]	392	382	374	372

[1]Preliminary, subject to revision.

[2]Breakdown may not add to totals due to rounding.

[3]A double-cropped acre is counted as 1 acre.

[4]Does not include cropland pasture or idle land not in Federal programs that is normally included in the total cropland base.

Sources: USDA/ERS, based on USDA/ERS, 1997a; USDA/NASS, 1998, 1999a, 2000a; 2004a; 2005; and unpublished data from USDA/FSA and USDA/NASS.

Cropland used for crops was at a record high of 387 million acres in 1949, when no acres were idled by Federal programs. In 1972, cropland used for crops was near a record low of 334 million acres as Federal programs idled 61 million acres.

Cropland used for crops climbed to 387 million acres in 1981, when Federal programs idled no cropland, and dropped to 333 million in 1983, when Federal program set-asides reached a historic peak of 78 million acres under the Payment-in-Kind (PIK) program.

The Federal Agricultural Improvement Act (FAIR) of 1996 eliminated all Federal acreage reduction programs other than the Conservation Reserve Program (CRP)

Between 1983 and 2002, cropland used for crops increased overall, while total acreage idled by Federal programs decreased to 34 million acres.

From 1997 to 2002, cropland used for crops declined from 349 to 340 million acres, while acreage idled under CRP increased by about 1 million acres. Between 2002 and 2004, acreage in CRP increased by an additional 1 million acres, while cropland used for crops declined by about 4 million acres.

The 14-million-acre drop in harvested cropland between 1997 and 2002 was coincident with a decrease in cultivated summer fallow and an increase in failed acres due to widespread drought. Crop failure occurred on 17 million acres, over 5 percent of the acreage planted, in 2002. This failed acreage was the largest since 1956. The use of summer fallow has been decreasing since the late 1960s, and stood at 15 million acres in 2002, down from 42 million acres in 1969. Four crops—corn for grain, soybeans, wheat, and hay—accounted for 80 percent of all crop acres harvested in 2002. The additional 17 "principal" crops accounted for another 15 percent of harvested area. Vegetables, fruits, nuts, melons, and all other crops accounted for 4.5 percent of crop area harvested in 2002.

Urbanization of Agricultural and Other Rural Land

Cropland conversion to urban uses is largely irreversible, so it is important to know the rate of conversion and how much of the loss is replaced from other land uses. Excessive loss of cropland to urban uses could lessen the production of food and fibre and the supply of rural amenities, such as open space, watershed protection, and rural lifestyles. A variety of Federal, State, local, and private programs address such concerns.

Although urban land constituted less than 3 percent of the U.S. land area in 2000, 79 percent of the population lived there. Even large percentage increases in urban area would amount to small decreases in rural area since it is so vast. The rate of expansion (by decade) of urban area has declined from 39 percent during the 1950s, to about 36 percent during the 1960s and the 1970s, and to 18 percent in the 1980s. According to the Census Bureau, urban area was 59 million acres in 2000, just 7 percent above the previous estimate for 1990 (DOC/BOC, 2002). However, the Census Bureau adopted a new definition of urban area for the 2000 Census, improving the precision of urban area measurement but making it more difficult to compare urban area before and after this year. If urban area for 1990 is recalculated using the 2000 definition, the Census Bureau's estimate falls to 51 million acres, implying a 15-percent increase in urban area from 1990 to 2000.

The National Resources Inventory (NRI), conducted by USDA's Natural Resources Conservation Service (NRCS) in cooperation with Iowa State University, is an alternative source for estimates of urban and rural areas. The NRI uses a consistent definition for built-up areas, though it differs from the definition used by the Census Bureau. According to the NRI, "developed land," which includes large and small urban and built-up areas as well as rural transportation land, totalled 107 million acres in 2002 in the contiguous United States. The NRI indicates that developed land increased by 14 million acres (19 percent) over 1982-92 and 21 million acres (24 percent) over 1992-2002 (USDA/NRCS, 2004).

Land converted to urban uses comes from several different major land uses. The NRI indicates that 20 percent of new developed land came from cropland between 1997 and 2002. Prime cropland— land that has the best combination of physical/chemical characteristics for agricultural production—is converted to developed uses at about the same rate (5 percent per year) as nonprime cropland (USDA/NRCS, 2003). About 21 percent of rural non-Federal land is prime and 26 percent of crop land converted to urban uses over 1997-2001 was prime.

Rural land, defined as all land that is not urban, contains rural residential land, consisting of houses and associated lots. Nonfarm rural residential area was estimated to be about 94 million acres in 2002, up from 56 million acres in 1980. The average rate of increase in rural residential land was 1.7 million acres per year from 1980 to 2002. Combining both rural and urban residential land, the total increase in residential area was about 2 million acres per year during this period.

Farm Real Estate Values

Farmland values rose rapidly during the 1970s and early 1980s, followed by a sharp decline during 1982-87, then a slow upward trend beginning in 1987. Since 1987, average farmland values in the Nation have increased 127 percent, from $599 per acre to $1,360 in January 2004. In real or inflation-adjusted terms (GDP deflator), however, this amounts to a 53-percent gain. It was not until January 1, 1995, that the average nominal value per acre surpassed the record high of $823 set in 1982. But the January 2004 average is still 8 percent below the 1982 average on a real (or inflation-adjusted) basis.

The 7.1-percent nominal increase in the national average value of agricultural real estate during 2004 marked the 17th consecutive increase since 1987. Over the previous 4 years, in particular, farm real estate values had increased substantially in all U.S. regions. Most notable is a 42-percent increase in the Lake States, versus 25 percent for all regions combined.

***Table:** Farm real estate values, by farm production region, January 1 for selected years*

Region	*1982*	*1987*	*2000*	*2001*	*2002*	*2003*	*2004*
$/acre							
Northeast	1,367	1,491	2,660	2,830	3,000	3,200	3,400
Lake States	1,234	707	1,560	1,700	1,870	2,010	2,220
Corn Belt	1,642	900	1,890	1,950	2,030	2,130	2,300
Northern Plains	547	331	535	556	576	594	632
Appalachian	1,083	1,004	1,990	2,120	2,250	2,370	2,500
Southeast	1,095	1,055	1,920	2,030	2,140	2,270	2,420
Delta States	1,135	757	1,270	1,330	1,390	1,490	1,550
Southern Plains	576	532	672	715	755	788	832
Mountain	325	257	448	471	500	523	550
Pacific	1,346	1,084	2,000	2,120	2,240	2,350	2,480
48 States	823	599	1,090	1,150	1,210	1,270	1,360

Source: National Agricultural Statistics Service, USDA.

The increases were widespread, with most States exhibiting increases for farm real estate, cropland, and pasture. As of January 2004, several Northeast States continued to record the highest average per-acre values for farm real estate, with Connecticut and Rhode Island exceeding $10,000 per acre.

These values reflect continued pressure from nonagricultural sources for conversion to residential or other urban-related uses. The high values in States such as California, Florida, Ohio, Illinois, Indiana, and North Carolina are the consequence of urban pressures, the production of high-value crops, or high soil fertility. The low real estate values for many States in the Northern Plains, Southern Plains, and Mountain regions can be attributed to large amounts of arid rangeland and less productive cropland. New Mexico, Wyoming, and Montana recorded the lowest average values per acre.

Cash Rents

Nearly a third of U.S. farmland is operated under some form of lease, according to the 2002 Census of Agriculture. The most common form of lease, the cash rental agreement, is a fixed payment negotiated before planting. Share rental agreements, by contrast, vary with the amount of product harvested. Under cash rental arrangements, the tenant bears all of the production and market-price risk; share rental arrangements divide production and market risks between tenant and landlord. Cash rents are generally considered a short-term indicator of the return to a landowner's investment. To tenants, though, cash rents are a major production expense and, like farm real estate values, have been increasing for a number of years.

Because rents reflect the income-earning capacity of the land, they vary widely across the country. Cropland rents tend to be highest in areas where higher-value crops are grown. The highest average cash rents in 2004 were reported for irrigated land in California, at $300 per acre. California produces large quantities of high-value speciality crops, vegetables, fruits, and nuts. Cropland most suitable for corn and soybean production, principally in the Midwest, also commands high rents. The highest rents for nonirrigated cropland in 2004 were reported as $126 per acre in both Illinois and Iowa.

During 2004, average cash rents for pasture varied from $37 per acre in Wisconsin to $1.70 per acre in New Mexico. States in the Appalachian, Delta, Southern Plains, and Pacific regions uniformly recorded increases from 2003.

Grazing Fees

Grazing fees for use of pasture or rangeland are also a form of cash rent, except that payment is based on "grazing units" rather than tracts of land (acres). A grazing unit is defined on an animal-unit-month (AUM) basis, which is one cow or cow-calf pair, or seven sheep/ goats, feeding for 1 month (NASS, 2005). Grazing fees on public lands administered by the Bureau of Land Management (BLM) of the U.S. Department of the Interior, and the Forest Service (FS) of the U.S. Department of Agriculture (USDA) are set by law. These fees vary annually according to a legislated formula, which links the fees to changes in the cost of production. As a result of the formula, 2005 grazing fees on public land were set at $1.79 per AUM. That marks the second consecutive year in which grazing fees were set above the statutory minimum $1.35 per AUM.

Grazing rates on privately owned nonirrigated land in 16 Western States averaged $14.30 per AUM in 2004. Rates ranged from $7 in Arizona and Oklahoma to $23 per AUM in Nebraska. Private grazing rates have trended upward since the early 1990s.

Factors Affecting Farm Real Estate Values

Traditionally, farmland value was based on its agricultural productivity. Particularly in the more rural areas of the Nation, where farmers still account for most farmland purchases, net returns to agricultural uses are the principal determinant of farmland value. Interest rates, capital investment in farm structures, and many other factors also influence productivity and thus the agricultural value of farm real estate. But today, many factors unrelated to productivity—including urban influence, government program payments, and rural amenities—contribute to the value of land in rural areas. In fact, these factors may be more important than productivity. High levels of direct government payments, which have occurred particularly since 1999, may have influenced farmland values in some regions.

Urban Influence

Farmland near cities has seen its value inflated by demand for conversion to nonfarm uses. As the U.S. population continues to grow and disperse, even primarily rural States such as Iowa are experiencing urban-related influences on farmland values. Commuters, who can now travel farther or even telecommute, are often willing to pay more than agricultural value in order to live in primarily rural areas. Other families develop hobby farms, second homes, or recreational structures in rural areas. In Iowa, for instance, there are now more nonfarmers living in rural areas than there are farmers. Other nonagricultural factors that may contribute value include the potential to concurrently use farmland for fee-based hunting, fee-based recreation, or wildlife viewing.

Nonfarm influences on agricultural real estate values have gained increased attention as interest in farmland preservation, suburban "sprawl," and habitat conservation has grown. Recent research indicates that nonfarm influence accounts for 25 percent of the market value of U.S. farmland. An ERS report recently addressed issues surrounding development of new houses, roads, and commercial buildings at the fringe of existing urban areas. This "sprawl" into the countryside can intersperse sometimes incompatible urban-related development with existing agriculture. Metropolitan Statistical Areas (MSAs) contain 20 percent of U.S. land area and 80 percent of U.S. population. The area also contained more than a third of all U.S. farms in 2003 and produced about a third of agricultural production value.

Direct Federal Payments

An array of government policies influence the income derived from farmland, and hence its value. Federal commodity and conservation programs are the most obvious. But also important are farm credit programs, State and local zoning regulations, habitat and species protection laws, infrastructure development (such as roads and dams), environmental regulations, and even property and income tax policy. Previous research has shown that capitalization of expected payments increases cropland values.

Also, the degree to which direct Federal payments are capitalized into cropland values depends upon the issuing program. If direct payments are capitalized into cropland values, as many theorize and some research has demonstrated, then a reduction in payments could signal a decline in cropland values and a loss of wealth for landowners. Further, ERS estimates that the degree to which direct payments (even from the same program) are capitalized into cropland values varies widely, with capitalization greatest in the Northern Plains. So from a policy perspective, the effect of program changes on cropland values would vary depending on the dominant program crop in a region.

Other Market-Related Factors

Interest rates, particularly inflation-adjusted ones, are especially important determinants of U.S. farmland values. As proxies for the discount rate, interest rates determine the current value of expected future earnings from land: for a given pattern of future earnings, higher (lower) interest rates imply lower (higher) land values. During much of the mid-to late 1970s, real interest rates were actually negative, providing a strong incentive to borrow money. Some of the borrowed money was used to purchase rapidly appreciating farmland. Conversely, real interest rates jumped from 1981 to 1985 when nominal interest rates increased rapidly just as expectations of future inflation were diminishing. The resulting increase in the real interest rate of mortgages has been cited as a cause of the slide in farmland values in the early and mid-1980s.

Inflation, lending policies of farm credit agencies and banks, and speculation also affect farmland values. And of course farmland values vary by site-specific characteristics like access to major highways, proximity to commodity and input markets, aesthetic appeal, and homesite potential.

Nonmarket Public Goods Provided by Farmland

Farmland also provides nonmonetary benefits. Until recently, these "rural amenity" benefits were supplied in such abundance that they were rarely acknowledged. But as the Nation becomes more urbanized, with the concomitant loss of farms and interspersion of urban-related activities, the decrease in those amenities has become a source of concern.

The nonmonetary benefits potentially reduced or eliminated by loss of farmland and open space include recreation opportunities, aesthetic enjoyment from viewing landscapes and wildlife, environmental quality, and nostalgia related to the historic and cultural significance of rural life. It is these "rural amenity" benefits that many farmland preservation programs seek to protect. A more extended discussion is available in AREI Chapter 5.6, Farmland Protection Programs, an ERS report on Rural Amenities, and current ERS activities examining farmland.

Land Ownership and Farm Structure

Ownership of U.S. Land

The land surface of the United States covers 2.3 billion acres. Private owners held 61 percent in 2002, the Federal Government 28 percent, State and local governments 9 percent, and Indian reservations 3 percent. Virtually all cropland is privately owned, as is three-fifths of grassland pasture and range and over half of forestland. Federal, State, and local government holdings consist primarily of forestland, rangeland, and other land. Most land in Federal ownership—largely in the West—is managed by the Department of the Interior (68 percent) and the Department of Agriculture (28 percent) (U.S. GSA, 2005).

Farm operators do not own all the land used in agriculture. According to the 1999 Agricultural Economics and Land Ownership Survey (AELOS), farmers held 58 percent of the land in farms in 1999 (USDA, 2001). These landowning farmers also made up 58 percent of the 3.4 million farmland owners.

Nonoperator landlords accounted for the remaining 42 percent of land in farms. Ninety-five percent of nonfarm landlords were individuals/families or partnerships. Of these unincorporated landlords, 55 percent were at least 65 years old. Many nonfarm landlords have a historic connection to farming. Among the people who have exited farming or inherited farmland since the number of farms peaked during the Great Depression, a number have retained ownership of some or all their land.

Farm Numbers, Farm Types, and Conservation Programs

The number of farms has declined dramatically since its peak of 6.8 million in 1935, with most of the decline occurring during the 1940s, 1950s, and 1960s. The decline in farm numbers has levelled off since the 1970s. By 2002, 2.1 million farms remained.

The remaining farms have a much larger average acreage, but averages mask differences among farms. Today's farms range from very small retirement and residential farms to industrialized operations with sales in the millions. Part of this diversity stems from the very low sales threshold ($1,000) necessary for an operation to qualify as a farm for statistical purposes.

One way to address the diversity of farms is to categorize them into more homogeneous groups. The farm typology developed by ERS identifies five groups of small family farms (sales less than $250,000): limited-resource, retirement, residential/lifestyle, farming-occupation/low-sales, and farming-occupation/high-sales.

The typology also includes large family farms, very large family farms, and nonfamily farms. In addition, very small farms (sales less than $10,000) make up more than half of all farms. Very small farms account for a particularly large share of farms in the limited-resource (72 percent), retirement (76 percent), and residential/lifestyle (76 percent) groups. Production, however, is concentrated among larger farms; small farms account for only 27 percent of the total value of production.

The smallness of most farms has implications for conservation and the environment. An ERS study found that smaller corn farms are less likely to use conservation tillage than are larger farms. The practice is more practical for larger farms because they have more acres over which to spread the cost of new or retrofitted equipment necessary to adopt conservation tillage. Small farms whose operators are retired or farm part-time are also less likely to adopt conservation tillage, possibly because of hesitancy to change familiar production practices. Small farms, however, participate widely in the Conservation Reserve Program (CRP) and the Wetlands Reserve Program (WRP).

Distribution of Conservation Program Payments by Type of Farm

These farms harvest most of the land planted to program commodities and therefore receive three-quarters of commodity program payments. However, CRP and WRP—the two major conservation programs—are targeted at particular types of land, not commodities. Since small farms own 70 percent of the land held by farms, they play a large role in natural resource and environmental policy. Retirement, residential/lifestyle, and low-sales farms account for nearly two-thirds of conservation payments and a similar share of the land farmers enrolled in the CRP and WRP. Participating farmers in each of the three groups tend to enroll large shares of their land in these programs: 46 percent of the land operated on retirement farms, 28 percent on residential/lifestyle farms, and 23 percent on low-sales farms. In contrast, enrollment ranges from 5 to 9 percent for participating high-sales, large, and very large farms.

Because their main job is off-farm, residential/lifestyle operators are limited in the amount of time they can spend farming. As a result, residential/lifestyle farmers find CRP and WRP attractive, since these programs require little time. Given their life-cycle position, many retired farmers have land available to put into conservation uses. The same forces may also be acting on low-sales operators, who average 57 years of age and may be scaling down their operations. If an off-farm job and advanced age are major determinants of land going into conservation uses, it may be relatively easy to get smaller farms to enrol land in the programs. Getting larger farms to enrol more of their land might require higher payments, if the opportunity cost of idling their land is higher.

Land Tenure

Farm operators leased 38 percent of their total farmland in 2002, down from 40 percent in 1997 and 43 percent in 1992, according to the census of agriculture. This decline may reflect increasing rental costs as parcels of land become smaller. Parcels of farmland available to rent tend to become subdivided with time due to division among heirs (Raup, 2003). Smaller parcels increase transaction costs to operators assembling land to expand their operations. Still, rented land as a share of total farmland is higher than the 35-percent rate that prevailed in the 1950s and 1960s.

About 38 percent of all farms rented land in 2003, 32 percent as part owners and 6 percent as tenants. Land leasing has changed from a way for beginning farmers to enter agriculture to a way for established farmers to access additional land.

Renting allows farms to expand without the debt and commitment of capital associated with ownership. In fact, about 17 percent of very large family farms are tenants, a larger percentage than in any other group. Conventional wisdom holds that farmland owners have a long-term interest in their land and thus are more likely than renters to adopt conservation practices. Soule and others (1999 and 2000) found this to be true among corn farmers, at least in the adoption of conservation practices that provide only long-term benefits, such as grassed waterways and strip cropping.

The situation was different for conservation tillage, which can increase profits in the short run by maintaining or increasing yields while reducing machinery, fuel, and labour costs. Cash-renters are less likely than owner-operators to use conservation tillage, but share-renters appear to act like owner-operators in adopting conservation tillage. Share-renters may have an incentive to adopt conservation tillage, if the landlord bears some of the costs that may increase under conservation tillage, such as herbicide expenditures. Share-landlords are also more likely to be involved in management decisions than cash-landlords, which may make share-renters act more like owners.

Conceptualising the Farming Systems Development Approach

The material in Part I is designed to provide background information that will be useful in placing the more technical discussion in Parts II and III on the implementation of the FSD approach in context.

Evolution of the FSD Approach

In the mid-1960's, there was little interaction between technical scientists (who were mostly on experiment stations) and social scientists (who tended to be concentrated in planning units). The Green Revolution was beginning to have a great deal of success in Asia and Latin America, being based on good climate (i.e., plenty of water) and soils; very homogeneous and favourable production environments; and the adoption of Improved varieties of wheat, maize, and rice that were very responsive to fertilizer. Improved inputs also were readily available and there was an accessible market for the products. However, in most of Sub-Saharan Africa, and certain parts of Latin America and Asia, there has been no Green Revolution. This is because climatic conditions are often not as favourable (i.e., too much or too little rainfall and limited amounts of irrigation), soils are generally poor, production environments are very heterogeneous and poor, and the input and output markets are poorly developed. Not surprisingly, there has been great difficulty in developing improved technologies that are attractive to farmers in such areas.

In Green Revolution areas, farmers were able to benefit from the improved technologies even if they did not do things quite right and the inputs they used were very divisible (e.g., they could use a little or a lot of improved fertilizer or seed). However, in areas with less hospitable environments (e.g., low rainfall areas that are found in many countries in Africa and Latin America), farmers have to do things exactly right if they are to benefit (e.g., planting on good soil moisture) and also they need lumpy inputs (e.g., control over traction). In addition, yield increases are not so good — they tend to involve incremental rather than major (i.e. revolutionary) changes in yields.

Thus, in the Green Revolution areas, because of the spectacular nature of the technology, experiment-station based technical scientists were very successful in their work. However, the lack of success in using a similar approach in poorer agricultural areas (i.e., with resource poor farmers), led to the evolution of the FSR approach, in which there is close cooperation between technical and social scientists. Work done with farmers in various countries in the 1960's and early 1970's revealed that these limited-resource farmers:

- Are rational (i.e., sensible) in the methods they use. For example, in Africa, there was little support for station-based research on mixed cropping until the early 1970's. although earlier farm-based research had revealed the rationality of the practice.
- Are natural experimenters. Obviously, the methods farmers naturally use will be those that appeal to them and are informal in nature in the sense that they are not usually amenable to formal statistical analysis.
- Understand the environment in which they operate rather complex farming systems, consisting of crops, livestock, and off-farm enterprises. In fact, it could be asserted that such

systems are often more complex than the specialized farming systems found in many high income countries. Unlike the case with limited resource farmers in low income countries, many of the constraints in specialized agriculture in high income countries can be broken or avoided through seeking advice and taking advantage of and receiving external help.

- Consequently, considerable respect developed for limited-resource farmers. The FSR approach evolved because of an increased awareness on the part of researchers that such farmers:
- Had a right to be involved in the technology development process, because they stood to gain or lose most from adoption of the technology.
- Could productively contribute to the development of appropriate improved technologies.

Therefore, the fundamental principle of FSR was that farmers could help in identifying the appropriate path to agricultural development. It is now recognized that limited resource farmers can be involved productively in all stages of the FSR approach. Farmers' participation at all stages relates in one way or the other to the selection, design, testing, and adoption of appropriate technologies.

FSR rapidly became popular and was strongly supported by many donor agencies. By the mid 1980s, about 250 medium-and long-term projects worldwide were carrying out FSD type work. As one of the major donors, USAID between 1978 and 1988 had funded 76 bilateral, regional, and centrally funded projects containing either a farming systems orientation or clearly focusing on farming systems type work. Forty-five of these were in Africa.

Now, many of those projects are being institutionalized within national programmes with considerable domestic financing. Thus, the FSR approach evolved primarily as a result of a lack of success in developing relevant Improved technologies.

The strong technical focus that characterised the evolution still persists to this day, although increasingly many, including FSD practitioners, are advocating that the approach can be used constructively in addressing not only technological solutions but also those relating to policy/support systems, a topic that will be considered further later in the manual.

In the mid-1980s, the Farm Management and Production Economics Service of FAO became actively involved in the farming systems movement and developed the FSD approach. FSD is based on the farm-household focus of FSR and emphasizes the central role the farmer plays in farming systems development. In addition, this approach incorporates an increased emphasis on the dissemination of improved farming systems based, in the process, on active involvement of those responsible for the policy/support systems. In the rest of the manual the term FSD will be used instead of FSR in order to avoid any possible confusion.

Comparing Station-Based Research and FSD

The literature frequently compares and contrasts station-based research and FSD. The tone in such comparisons often implies that they are substitutes for each other. However, both station based research and farm-level FSD are needed. This is because they focus on different things that are complementary to each other.

A major difference between them is the following:

- On the experiment station, applied research is usually undertaken, in which new technologies are created.
- FSD, on the other hand, concentrates mainly on adaptive research, which involves helping to adjust technologies to specific environmental conditions. FSD also helps feed back information about future priorities for applied research to experiment stations.

Some of the major differences between station-based research and FSD and helps indicate why both are necessary, particularly in areas where the Green Revolution has not occurred. In such areas, greater attention will need to be paid to adaptive research, if relevant improved technologies are to be developed and adopted by farmers.

Obviously, strong linkages between these various contributors are of crucial importance. The FSD approach, involving working directly with the farmers, is a more 'bottom-up' or 'micro to macro, orientation, compared with the more 'topdown, or 'macro to micro' orientation of research work or planning exercises that start at the experiment station or the upper levels of planning ministries.

Part 'B' indicates that FSD can help in strengthening 'bottom up' linkages amongst the various groups. In doing so, it helps in the process of improving productivity but does not result in a product by itself. (i.e., facilitates a process rather than producing a product).

As mentioned above, relevant improved practices/technologies and policy/support systems are needed to bring about increased agricultural productivity. Decisions on which strategy is to be emphasized to increase agricultural productivity will depend on circumstances.

For example, if no technology is available to ensure an economic return to fertilizer, there will be little value in concentrating on developing an input distribution system that makes fertilizer available to farmers. Therefore' the complementary nature of the relationship between developing and disseminating relevant improved technologies and policy/support systems needs to be constantly borne in mind, if the agricultural development process is to proceed efficiently.

In the above discussion, it is apparent that the FSD approach has a broader focus than simply addressing technological issues— the primary focus of early FSR activities. The FSD approach, can help also in facilitating linkages not only between farmers and station-based researchers but also with other 'actors,, including those responsible for designing and implementing the policy/support system.

In fact, most work in the farming systems arena to date has concentrated on the technology thrust or farming systems activities relating to FSD. This manual concentrates most attention on the research or technology thrust of FSD although linkages with other 'actors' besides farmers and station-based researchers are discussed to some extent.

General Approaches to the Farming Systems Approach to Development

Introduction to FSD

Details concerning the characteristics and methodology of the FSD approach, as developed by the Farm Management and Production Economics Service of FAO, are available elsewhere. Thus, the following discussion represents only a brief summary. The primary objective of FSD is to improve the well-being of individual farming families by increasing the overall productivity of the farming system in the context of both the private and societal goals, given the constraints and potentials imposed by the factors that determine the existing farming system. It is based on the development principles of improving productivity, increasing profitability, ensuring sustainability, and guaranteeing an equitable distribution of the results of production,

The farm household is the principal system and focus of FSD and consists of three basic subsystems, which are closely interlinked and interactive: the household, the farm, and off-farm activities. The two major categories of activities or thrusts of FSD, both of which involve intensive interaction with farmers? are:

- Farming Systems Analysis. This involves studying, together with the farmers the natural (i.e., technical) and socio-economic (i.e., human) environments in which farm households operate. The aim is to identify the constraints limiting farm productivity and production and hindering improvement in the welfare of the farm households themselves. Potential solutions to these problems are identified, and the results of this analysis, formulated as suggestions for further action then are passed on to the relevant 'actors'. These could include researchers, extension and support service staff, and/or policy makers.
- Farming Systems Planning, Monitoring, and Evaluation. These involve testing, monitoring, and evaluating improvements on-farm, with the direct involvement of farmers. Examples of such activities include those regarding proposed technological improvements, proposed revisions in farm plans, and improvements in support services and farm-level impact of proposed policy changes. Those improvements thought to be potentially useful then are disseminated to other farmers via the extension service. Prior to this stage, it may be necessary to negotiate adjustments in the policy/support services that will facilitate the dissemination and adoption of the improvement. After dissemination, it is of primary importance to monitor and evaluate the adoption rate of the proposed improvements by the farming community. This provides an indication of the actual value of the improvement being disseminated.
- Many individuals in the last 15 to 20 years have written extensively to clarity the concepts of the farming systems, particularly with respect to the development of relevant improved technologies. There are many ways of presenting a diagram involving the application of farming systems

methodology to research (i.e., technology development). One such way of emphasizing the technical research aspects of FSD. There are four fundamental stages in the FSD approach.

- The Descriptive or Diagnostic Stage, in which the actual farming system is examined in the context of the 'total, environment — to identify constraints farmers face and to determine the potential flexibility in the farming system in terms of timing, unused resources, etc. An effort also is made to understand the goals and motivation of farmers that may affect their efforts to improve the farming system,

The following sequence of activities is usually undertaken:

  o After deciding on the location of the work, start-up activities include reviewing secondary sources of information, making the necessary contacts, assembling the professional team, and making logistical arrangements,

  o Farming families are classified tentatively into homogeneous groups. The farming families within each group or domain usually practice the same farming system(s), face the same constraints, and have the same potential solution(s) to their problems. This tentative classification can be modified as additional information becomes available.

  o An informal exploratory diagnosis or survey is undertaken to obtain a qualitative understanding of the determinants of the existing farming systems (i.e., both biophysical and socioeconomic). At the same time, efforts are made with farmers to ascertain the constraints, flexibility, and potential opportunities in the farming systems they are using currently.

  * Sometimes a verification survey involving a structured formal survey, suitable for statistical analysis, is undertaken to quantitatively confirm insights obtained in the exploratory diagnosis. It also can be used to provide a database for farm and development planning, policy analysis, monitoring, and evaluation.

- The Design Stage, in which a range of strategies is identified that are thought to be relevant in dealing with the constraints determined in the descriptive or diagnostic stage.

The usual sequence of activities involves the following:

- o Together with the farmers, the constraints are ranked according to their severity, and potential solutions are identified after determining what flexibility exists in the farming systems currently practiced. Information for designing such strategies (i.e., particularly relating to technological issues) comes from experiment station work, researcher managed and researcher implemented (RMRI) type work on farmers' fields, and from other farmers.
- o An evaluation is made of the proposed solutions before putting them into practice. These solutions can be technological or institutional in nature. A number of analytical techniques can be used to evaluate the potential technical feasibility? economic viability, social acceptability, and ecological sustainability of the proposed solutions before they are put into practice. One or more of the proposed solutions are selected for actual evaluation on-farm.

- The Testing and Implementation Stage, in which one or more promising strategies, arising from the design stage, are examined and evaluated under term conditions to determine their suitability for producing desirable and acceptable changes in the existing farming system.

Activities at this stage involve the following:

- o On-farm testing or evaluation with farmers determining how well the potential improvements fit into the system, whether or not they are acceptable to farming households, and what modification may be needed to make them acceptable. In terms of technological issues, this stage conventionally has consisted of two steps:
  - — Researcher managed but farmer implemented (RMFI) tests to establish whether previously determined technical relationships are altered by farmers' management of non-treatment variables.
  - — Farmer managed and implemented (FMFI) tests, when the team is confident that technical relationships will hold but needs to evaluate the proposed technologies under local socio-economic circumstances. In addition to the on-farm evaluation of the proposed

technologies, evaluations also can be made of the proposed farm plans and of proposed changes in support systems and/or policies.

- o Positive results of such evaluations provide justification for the FSD team to advocate further action. The tested technologies can be disseminated through the extension service to other similar farming households, The same applies to the farm plans that have been tested successfully. Favourable test results for proposed changes in support services and/or policies provide valuable farm-level information on necessary programme or policy adjustments (e.g., in extension, marketing inputs and products, pricing policy, credit, etc).

- The Dissemination and Impact Evaluation Stage, in which the strategies that were identified and screened during the design and testing stages are extended to farmers. In terms of activities at this stage, impact/adoption studies can be very important. These potentially can be very useful, not only in giving some idea of the impact of, for example, agricultural research, but also in giving some idea of future priorities for agricultural researchers and indicating what adjustments are required in the policy/support systems to ensure better rates of adoption. Thus, through monitoring and evaluating the impact and rate of adoption of the changes that have been implemented and proposed/tested earlier, it is potentially possible to provide some indication of further desirable activities by both researchers and planners.

There are often no clear boundaries between the various stages, Design activity, for example, may begin before the descriptive and diagnostic stages and may continue into the testing stage, as promising alternatives emerge from RMFI trials.

Similarly, testing by farmers may mark the beginning of dissemination activities. Also, going through all stages may not always be necessary. FSD team confidence in transferability during design/ planning activities can sometimes mean going straight to FMFI work or even to recommendation/dissemination activities.

Thus, the process of FSD is recognized as being dynamic and iterative, with linkages in both directions between farmers, researchers, extension staff and policy/support service staff, The iterative characteristic can improve the efficiency of the development process by providing a means of identifying and fine tuning improved technologies for a specific location-that is, climatic situation, soil type, and/or farmer resource base.

Approaches within Farming Systems Research

Because FSD has its roots in farming systems research, the following comments on approaches within FSR are applicable to the technology research activities of FSD. The term FSD often is used loosely. An activity is considered to be a legitimate part of the technology research function of FSD if the following situations hold:

- The whole farm is viewed as a system.
- Research is conducted with a recognition and emphasis on the choice of priorities that reflect the whole farm.
- Research on a farm sub-system is legitimate FSD, provided the connections with other sub-systems are recognized and taken into account.
- Evaluation of research results explicitly takes into account linkages between sub-systems.
- As long as the concept of the whole farm and its environment is preserved, not all the factors determining the farming system need to be considered as variables — some may be treated as parameters or constants.

A distinctive feature of FSR, or rather FSD, is that a systems perspective is constantly borne in mind. Within this framework, one can look at either a small or a large number of variables in the system. In general, because of the complexities of simultaneously handling a large number of variables, most FSR programmes have tended to limit the number of variables they study and to regard the other factors that influence the farming system as parameters or constants.

Three general approaches to FSR or FSD with a technology generation focus, have been defined relating to the number of variables being investigated. These approaches can be described as FSD 'with a pre-determined focus', 'in the small', and 'in the large', In a sense, these types represent an evolution in FSD as techniques have been developed to handle progressively more complex situations— defined as involving a higher ratio of variables to parameters, The progression, which is illustrated, where it has been expanded to four phases, can be articulated as follows:

- In its early days, FSR (i.e., FSD) type activities concentrated on how farm yields of particular crops could be raised. This form of on-farm research stood in sharp contrast to earlier multi-locational on-farm trials, in which only the effects of the physical (i.e., natural or technical) environment were tested, Even though this FSD approach did involve the inclusion of socio-economic elements and, hence, had a farming systems perspective, it was done with a 'predetermined focus' on the productivity of a particular commodity. Thus, this approach involves looking at one facet of an enterprise or one specific enterprise and identifying improvements within that focus that are compatible with the whole farming system. With the creation of distinct FSD teams — in contrast to an on-farm testing component located within each station based commodity team — FSD activities generally have evolved beyond this level. However, it is important to recognise that this type of activity still is a legitimate part of any FSD in the sense that it can provide a very useful input into the activities of commodity research teams.
- In the pursuit of greater farmer participation, research shifted to addressing farmers' articulated problems directly. Still, farmers' problems were very much con fined to crop production, although links with other components of the whole farm system were studied.

  An alternative name for this 'farming systems in the small' focus could be 'farming systems with a whole farm problem focus', Many FSD efforts currently tend to be at this point, However, currently, on-farm research projects that seek to improve whole farm performance through manipulation of the linkages between all enterprises are :

  — Looking further ahead and to enable addressing issues of sustainable agriculture effectively, concerns for maximizing the economic and biological performance of enterprises will be modified by concerns relating to ensuring sustainability, including the rehabilitation and regeneration of natural resource systems. This natural resource system dimension to sustainable agriculture can be thought of as a type of 'farming systems in the large, or 'farming systems with a natural resource systems focus',

— Similarly, attention to ecological sustainability will need expanding to include the many off-farm and non-farm activities that make up sustainable livelihoods. This also can be thought of as a type of 'farming systems in the large, or 'farming systems with a livelihood systems focus'.

These notions, particularly the last two, have been largely unexplored until recently, when techniques have evolved to handle complex situations in which the ratio of variables to parameters is high. Formal modelling techniques by themselves generally have limited value in addressing such situations because of the complexity of the relationships and the degree of understanding that is required initially to develop a realistic model. However, the major breakthrough that has occurred is the ability to pursue a less formal or more informal modelling approach through application of farmer participatory type techniques that, in essence, involve using farmers' minds as computers!

Fsd with A 'Pre-determined Focus' on Specific Technology

Researchers in station-based programmes often want FSD workers to evaluate specific technologies they have developed to see how they fit in with constraints of the farming system currently used by farmers. This type of work is referred to as FSD with a 'pre-determined focus' or 'predetermined technology focus,' In Botswana, FSD teams and participating farmers have assisted station programmes in testing cowpea varieties, evaluating the desirability of hilling groundnuts, ascertaining whether the Dutch hoe can help with the weeding operation, and so forth.

This type of testing in FSD is not analogous to typical technical transfer. Rather than testing fully finished products, FSD assists technology generating programmes, usually based on research stations. Often, the types of technology tested in this manner have not been identified as having high leverage power in the system.

Leverage power refers to the ability of a technology change to have a major impact on the performance of the farming system. However, these technologies do fit possible needs and asking farmers to participate in their development is worthwhile, It is desirable to test technologies as early as possible in their development, This is to facilitate making modifications and to minimize waste of research resources. It also may accelerate the dissemination of suitable technology, Such technology, even if not fully appropriate, may sometimes stimulate thinking among farmers if they are not accustomed to these types of interventions. Feedback in these instances can contribute to new innovations in technology generation.

FSD with a 'Pre-determined Focus' on a Subject

A great deal of FSD work, particularly in its earlier years, was not focused on testing specific technology as much as a pre-determined focus on a subject matter area. FSD within the CGIAR institutions such as IRRI, CIMMYT, and ICRISAT have focused on issues related to production of their respective commodity crops. ILCA focused on issues in livestock production, ICRAF in agroforestry' etc. Some FSD teams within the national agricultural research systems (NARS) also worked under limited mandates to deal with particular enterprises; commodities; or other subject areas.

The judgements of farmers, the end-users of products developed by these organizations, were foremost in these FSD activities. However, the parent institution with its technologygeneration activities would be a primary client in these instances.

FSD with a 'Whole-farm Problem Perspective, or 'FSD in the Small'

An FSD interdisciplinary team conducted a diagnostic survey and determined that labour for satisfactory weed control in field crops is a major constraint to farm production and farm family well-being. The survey also showed that due to population pressures, farmers were no longer able to shift field sites as in the past and that deteriorating soil fertility also was becoming a major issue.

However, during the course of farmers participating in testing possible solutions to these problems, discussions between the FSD team and farmers identified the potential for intensified poultry production as an additional means of improving farm income. This resulted, in addition, in collaborative testing in the area of intensive poultry production and, as a result, exemplifies FSD with a 'whole-farm problem perspective.'

The major breakthrough that has occurred is through supplementing the rapid rural appraisal (RRA) techniques, which have been available for some time with the participatory rural appraisal (PRA) techniques, which have developed rapidly in recent years. To date, the first two approaches listed above have been used. Although

FSD 'in the small' arrives at a focus within the system in the course of diagnosis, FSD 'with. pre determined focus' moves into the system to research an enterprise or one facet of an enterprise looking for improvements within that focus that are compatible with the whole farming system.

FSD with A 'Natural-resource Management Focus

A non-governmental organization (NGO) has adopted, as one of its aims, the conservation of the natural resource base within areas of agricultural production. This NGO sponsors FSD activities in a village in which participating farmers look at current and prospective problems in natural resource conservation and management. They evaluate potential solutions to these problems at the same time issues in agricultural production and family income are considered. The FSD team from the NGO and farmer participants discovered a number of incongruencies between issues of conservation and present day agricultural production. They included the following:

- Natural-resource conservation analyses focus on the landscape (e.g., possibly a watershed), which in some instances does not even correspond to a village or to any other political entity. On the other hand, agricultural production analyses tend to focus on individual farms, units within farms, or smaller interacting networks of farms.
- Conservation invariably takes a long-term perspective of problems and solutions, whereas production analyses consider short-term returns and immediate survival issues.
- Natural resource conservation often involves community or ultra-community efforts and goals, which might conflict in terms of labour or other requirements, with individual farm production activities.

Integrating these apparently opposing sets of objectives for natural resource conservation and production poses a major challenge for FSD. Participating farmers recognize many of the conservation issues and trends and some express concern for the future.

In some cases, traditional techniques for conservation such as gully amelioration have been identified and are being promoted, but other issues remain unresolved. Some of the constraints encountered today or anticipated in the future were unknown to farmers in the past. Solutions therefore, are not known, and innovative and problem-solving thinking is required. However, given farmers' intimate knowledge of their current and past production environment, their input in designing appropriate strategies for the future based on an analysis of what has happened in the past is indispensable. This is clearly a fertile area for active collaboration between on-farm and on-station work: a give and take between farmers, scientists, and policy makers.

Some Characteristics of FSD

Some of the major characteristics of FSD are as follows:

- Farmer Centre Stage. The farmer, as the consumer of the improved technologies, is in the centre of the stage. This provides an opportunity for researchers to learn from the farmer, enables him or her to have an input into the research process, and ensures that criteria relevant to him or her are used in evaluating proposed technologies. For the farming family, evaluation criteria (i.e., for the adoption of the improved technologies) can be divided into the following groups:
    - Necessary conditions determine whether the farming family would be able to adopt the improved practices. Such conditions include technical feasibility, social acceptability, and compatibility with external institutions — that is, support systems.
    - Sufficient conditions determine whether the farmer would be willing to adopt the improved practices. Obviously, the necessary conditions will be influential in determining this willingness. Sufficient conditions include the compatibility of the improved practices with the goal(s) (e.g., food self-sufficiency, profit maximization, risk minimization, etc.) of the farming family; the resources they have access to; and the farming system they now practice.
- Work with Representative Farmers. Although the input of farmers is critically important in FSD it is impossible for FSD teams to work with all farmers. Therefore, a few are selected that are thought to be representative of all farmers. Because there are many types of farmers— with differences in the products they produce, the resources they possess, and the problems they face — it is necessary to put farmers with similar characteristics in the same group (i.e., stratum or

recommendation domain). A small sample of farmers in each of these groups then is selected to work with the FSD team. If the grouping and selection of farmers are carried out correctly, then the results obtained by these representative farmers should be achievable by other farmers with similar characteristics, when the technologies are extended to other farmers by extension staff.

- Involves an Interdisciplinary Approach. As indicated earlier, farmers have complex farming systems. As a result, changes in one part of the farming system may have a good or bad impact on another part of the farming system. For example, in some areas, it is not realistic, or indeed desirable, to try and increase crop productivity without considering livestock at all,' Also, attaining a successful solution to a technical problem that has been identified will depend on whether the farming family has the labour and money to adopt (i.e., an economic issue) and any sociological reasons preventing them from adopting. Therefore, to address the wide range of farming systems' issues, FSD teams generally consist of representatives of a number of disciplines — usually agronomists, animal scientists, and agricultural economists and sometimes sociologists. As a result, an interdisciplinary approach, that is a number of disciplines working together on the same problem, has to be used to solve the problems of farmers.
- Dynamic and Iterative Approach. Because of the dynamic nature of agriculture, research is a never-ending process. Farmers always face problems to varying degrees in their farming operation, and many of these problems can be solved through research. Sometimes the solutions suggested as a result of research don't work and need to be modified. The dotted lines indicate that such failures require moving back to an earlier stage in the FSD process and repeating one or more of the steps in the process. This makes FSD an iterative as well as a dynamic approach.
- Complementary to Station-based Research. In the context of technology choice and development, the role of FSD (i.e., utilising a systems approach) is seen to be complementary to technical component research (i.e., which primarily employs a reductionist approach), most of which is undertaken on

experiment stations and is usually commodity-based, With respect to such research, FSD has three roles:

- o To look at recognized farming systems and the stock of materials and techniques accumulated from station-based research, so as to be able to choose technical solutions to problems that have been identified, On-farm experimentation then adapts chosen solutions to the local situation. This is a mobilizing and adaptive role, shaping the product for an identified market.
- o To pass back unsolved technical problems, important to the system, to the appropriate commodity research team on the experiment station, This role is one of identifying and helping to prioritize the agenda for technical research.
- o To link with farmer clients and extension staff in local farm situations, drawing both farmers and extension workers and other relevant 'actors' into the technology-generation process.

Two examples indicating the perils of not exploiting the complementarily between on farm and station-based research are given in Boxes 3.6 and 3,7. Similar linkages could be established with planning and development agencies concerning proposed policy/support programme changes. Unfortunately, these are not yet strongly developed in most countries.

FSD with A 'Livelihood-systems Focus

There are, to date, relatively few good examples of applying FSD with a 'livelihood systems focus.' One organization that in recent years has expended considerable effort in integrating 'off farm' with 'on-farm' activities in a sustainable manner has been ICLARM which has concentrated on integrating aquaculture with crops and livestock.

Certainly integration of 'on-farm' and 'non-farm' activities deserve much greater emphasis than it has to date, especially as increasing pressures develop on the natural resource base. 'Non-farm' activities car provide an outlet for surplus labour and contribute to the attainment of sustainable livelihoods.

In many areas (e.g., semi-arid parts of West Africa), particularly where agricultural activities are seasonal in nature, farmers have always recognized the complementarily between farm and non-farm activities, through emphasizing the latter particularly during slack periods in the agricultural cycle, Many of these jobs are in the non-formal sector and involve micro-enterprises, also an area that needs much greater emphasis than it has to date.

Technology that Works Under Experimental Conditions will not Always be Relevant on-farm

In Botswana, precipitation is low and variable, creating erratic conditions for crop production. Under these conditions, farmers minimize risk by adopting an extensive low-input cropping system that is moderately productive under favourable conditions but minimizes losses when rainfall fails. Under research-station management, an intensive system for maize increased average yields and dramatically improved yield stability. The intensive system consisted of controlled pathways for planting with minimal tillage of the soil and water concentration onto the plant zone. It is the water concentration that stabilizes maize yields. Weed control was as used needed but weed pressures were minimal when crops were established early.

This technology failed when it was tested under farmer management. It failed because critical resources are not available on the farm at the right time to meet the schedule requirements of the new technology. The intensive system requires that soil preparation be finished before the rainy season begins and that planting be carried out on the first rain(s) of the year. Meanwhile, draught animals to do this work are kept near watering points at distant cattle posts until rains have filled shallow wells and ponds near the herds. Farmers and their draught animals cannot plant early in the season, and when they do not, weed growth overruns the controlled pathways. The Ministry of Agriculture made a fitting decision to conduct farmer-managed tests before recommending this system to farmers based on station results alone.

A Systems or Reductionist Approach?

Is this the only choice? It might be popular to think in terms of the one or the other; systems, approaches or reductionism, each with its characteristics and methodology. Should one 'divide and conquer' or accept that 'the whole is greater than the sum of its parts"? A distinction is made between science and art but also it is reflected in divergent schools of thinking within some disciplines. However, in FSD, this rigid dichotomy of perspective is neither necessary or feasible.

FSD recognizes that farming systems are highly complex and not fully reducible. They are too complex for human intuition to deal with all the factors involved but, at the same time, consist of such a fuzzy web of interactions that currently available analytical tools are not powerful enough to describe them.

Over the years, science in agriculture has attempted to model the factors and relationships within a farm operating in an agro-ecological situation. The purpose of this modelling invariably has been to engineer optimal solutions or strategies. An important limitation to this approach is that optimization has been for the research model and not for the farm and the farming family itself. A larger purposeful framework comprising a farm value system and related decision making patterns cannot be incorporated so readily into this research model.

FSD seeks to synthesize human intuition and judgement with analytical science. To accomplish this, FSD uses farmer perspectives on issues, problems, etc. Even when these are not completely exact, they help create holistic or logical frameworks within which technical alternatives can be screened or engineered. FSD also seeks to render the holistic framework as explicit as possible. Farmers and FSD staff seek to differentiate uncertainty in the system that is due to randomness (i.e., rainfall probabilities) from that due to lack of experience (i.e., ignorance about new technology) from that due to vagueness (i.e., typically, decision making patterns). By integrating relevant scientific information, hands-on experiences, and intuitive managerial skills, FSD can be effective in evaluating scenarios, even those involving future changes in the farming environment.

Within this framework, numerous scientific information needs will be recognized. Within this framework, the need is to obtain rather unequivocal answers to the specific and well defined questions. Reductionist's techniques are better suited for this purpose, especially if the question pertains to objects in the system rather than human behaviour, which involves human intuition skills. The key to successfully designing this type of focused research is the incorporation of different perspectives through interdisciplinary teamwork and participation of farmer managers. Thus, 'rigour versus relevancy' need not be a dilemma in FSD, On the contrary, rigour and relevancy are both possible and essential, if solutions are to be found in time for the many pressing problems that confront farming systems in a rapidly changing world.

The systems analysis employed in FSD should do the following:

- Pay particular attention to formulating problems within the holistic framework.
- Review each result within the general systems framework and never exclude an alternative without having done this. Sometimes, the best alternative in a systems framework is tar from optimal when evaluated separately.
- Set forth hypotheses early, so that all 'actors' with their perspectives can participate in reformulating these questions. After all, it is these questions more than 'responses, that set systems approaches apart.
- Avoid overemphasizing mathematical representation at the earliest stages.
- Treat even uncertainty as explicitly as possible.
- Sub-optimize with care.

Thus, there is a complementarity between the reductionist and systems approaches. This is why it was argued earlier that the FSD approach is complementary to station-based research, which usually employs a reductionist approach. The holistic approach in FSD is needed to ensure that experiment stations develop relevant technologies and needs to exploit the power of the reductionist approach, which can be a very efficient way of developing new technologies, once the priorities have been ascertained correctly. The results of this potentially relevant technological work can then be given to the FSD team for on-farm evaluation.

Appropriate Technology' will not Always be Relevant on-farm

Appropriate technology is characterized as low cost, technologically simple, labour intensive, making use of local materials and existing methods of production, and responding to identified constraints in the farming system. FSD generally seeks technologically appropriate solutions to problems but these should not be promoted automatically without on-farm assessments. The Bolivian altiplano is a hostile environment for agricultural production. The zone is dry and cold with nightly frosts occurring at any time of year. In earlier periods, farmers of this area traded products of their system for various food products of the lowlands. In recent times, their access to other markets has partially been broken off. Land degradation and pressures on use of the land have further reduced the ability of farmers to maintain their traditional crop and livestock production strategies. Protected horticultural systems were identified by several NGOs as appropriate ways to solve problems in this system. For the most part, these systems were promoted directly without an objective assessment including testing under farmers' conditions. Initial glasshouses were soon found to have difficulties with high insolation costs and were replaced by adobe structures covered with plastic. But many other problems were not addressed or proved unsolvable: guaranteed water supplies, local and distant markets for excess production, nutritional benefits of products, and so forth.

An important finding of a study of this large and costly effort is that promoters of development have not taken experimentation seriously enough. The value of farmer and expert involvement in design, testing, and dissemination is that irrelevant or incorrect designs and management systems can be screened out or corrected.

FSD in the Context of other Agricultural Institutions

Work relating to FSD might be incorporated as a working tool for several different types of institutions. These could include such diverse institutions as agricultural research, extension, community development (particularly in regards to NGOs), or even ministerial or national units interested in policy design and assessment.

FSD thus can he used to focus on family income. technology design, dissemination, and adoption: or designing solutions to the problems associated with a deteriorating natural resource base in agricultural areas. At another level of viewing the institutional setting of FSD, it is apparent that not only some disciplines but also various technical and sociological approaches and methods, are evolving to better deal with systems analysis. Take, for examples, the use of Geographic Information Systems (GIS) as a tool integrating large amounts of data into a single framework useful for technical as well as sociological and economic analyses. On the other hand, community development and policy reform programmes use increasingly sophisticated techniques (i.e., tools or methods) that contribute to decentralized planning, grass-roots participation, and empowerment in assessing problems and designing solutions. The emphasis in many areas is on developing integrated approaches and system type analytical methods. With the possibility of increasing convergence occurring in some of the methods and, in some cases, objectives, the unique contributions by FSD will be brought into question by some. And, in fact, the institutional role or mandate of FSD will vary. FSD in one setting may focus Iargely on trials and farmer testing of new technology. In another setting, a greater emphasis may be placed on surveys, sociological or information related to the financial picture of the whole farm, and so forth, In the larger picture, none of this matters too much. But it should be remembered that FSD does have certain fundamental characteristics. These include:

- Iteration, such that no single proposal should be considered definitive.
- Integration, usually involving multiple disciplines and explicit building on managerial perspectives and managerial values.
- A consistent orientation towards problem solving and an attitude that no perspective on this problem should be discounted.

FSD also must focus on appropriate linkages with other development support programs and institutions. In this regard, FSD emphasizes a strongly integrative and integrating philosophy, not just for disciplinary and farming perspectives within FSD itself; hut integrating strengths of personnel, resources, information, and location of the FSD programme with other research and development programmes working in the same area or domain. A critical element of this inter-institutional integration is a logical division of labour for the problem-solving task, This invariably will encourage collaboration in the development of work plans both for short-term as well as medium-and long-term objectives.

Challenges Facing FSD

A number of challenges face FSD if it is to continue playing a significant role in facilitating the process of agricultural development. Some of these are:

- Better Incorporation of Farmers, Incorporating farmers into the research process was one of the most important principles underlying the evolution of the farming systems approach. The basic justification for this was that farmers could improve the efficiency of the research process. At a minimum, it was argued they could prevent the use of research resources on types of technologies that they would be unlikely to adopt, while even more Importantly, it was anticipated they could, in their own right, contribute positively to the research process.

Unfortunately, farming systems workers often have not sufficiently recognized this positive and interactive contribution of farmers. Consequently, this spawned the farmer participatory research (FPR) movement. The problem of not utilizing the farmer sufficiently in FSD activities is not due to inherent deficiencies in FSD itself but rather with the way in which it has been applied, Techniques of FPR or PRA need to be incorporated into FSD, a subject which will be discussed later.

- Continued Evolution of FSD. FSD is relatively new. Therefore, the methodology is still evolving and, as a result, universally accepted 'standard texts' on the 'nuts and bolts, of how to do it are still to emerge. Related to the methodology issue is that time and cost-efficient (i.e., money and people) methods for undertaking FSD still need further development. This is important because of the limited resources available for undertaking research, both on-station and on-farm.
- Greater Incorporation of the Policy/Support System Perspective. As has been stressed already' the farming systems approaches implemented to date have focused mainly on the technology dimension (i.e., FSR). However, a basic principle of the FSD approach is that the farming systems perspective is critically important in formulating and adapting policy/support systems in ways that will facilitate and accelerate the agricultural development process, Once again, discussions relating to the policy/support system are scattered throughout the manual, and it is highlighted as an essential component of FSD. Failure to incorporate this dimension in FSD activities is likely to have an impact analogous to playing soccer on one leg, Nevertheless, because of the lack of much proven experience and documented material on the farming system perspective with respect to the policy/support system, this manual does not deal with this dimension to the extent that would be desirable. Hopefully, this lack of experience will be rectified in the fairly near future. An example of the potential value of using a policy/support perspective in FSD is given in Box 3.8.
- Incorporating Equity Issues — Intra and lnter-Generational. FSD tries to help the farmer with the problems he or she has identified. Of course, the reason for this is the necessity to introduce an intervention in which they are interested, It its

likely that these 'felt, problems of farmers are likely to have a short-run focus (i.e., particularly when the farmer is operating very close to the survival level). Also, it is possible that helping to solve the individual farmers' problems creates others for the society as a whole. Although the example given in Box 3.9 pertains to inter-household relationships, equity considerations within a particular generation also apply to what is happening within farming families. As will be discussed later, it is wrong to assume that all farming households or families operate in such a way that distribution of effort and benefits is equitable. For a positive example of the benefits of understanding intra household relationships. When designing a technology to help farmers increase their productivity, consideration must be given to the possible long-term effects (e.g., decreasing the amount of productive land available) of that intervention. Therefore, if farming system (FS) workers are not careful, their work can result in creating two types of inequalities, that is, helping some farming households— or even certain individuals within those households — at the expense of others and/or reducing the quantity and quality of land that can be productively farmed by future generations. Avoiding the development of such inequities constitutes a major challenge to FSD teams who need techniques to be able to screen out proposed technologies and policy/support systems that would encourage or exacerbate such trends. Once again, issues relating to equity matters, although generally recognized as important, still need attention in farming systems type activities. This will require continued evolution of FSD towards the FS 'in the large' approach mentioned earlier.

- Assessing Agricultural Research Impact. Related to the research resource issue is the importance of devoting some effort to assessing the impact of the research process-something that often has been done inadequately. More attention needs to be paid to adoption/diffusion studies. Such studies, of course, are the best measure of the impact of the agricultural development process, Therefore, the credit for favourable results from such studies cannot be allocated to FSD activities alone. However, FSD teams, because of their on-farm location and pivotal linkages with other actors, are often in the best position to take a leadership role in their execution. Also, as suggested earlier, such studies can and should be used for other purposes, such as feeding back priorities for further research and providing evidence for adjustments in the policy/support system to encourage greater adoption,

- Improving Credibility of FSD. Establishing credibility for FSD-related activities is a major challenge and is necessary to ensure that some of the limited research resources always will be allocated to them, As argued earlier, FSD staff constitute only one set of the 'actors' in the agricultural development process. Additionally, FSD helps facilitate a process and does not result in a product by itself. Thus, FSD cannot claim sole credit for any technologies developed for, or adopted by, farmers. However, it achieves credibility through its linkages and cooperative efforts with other 'actors' in the agricultural development process. These 'intermediate products' need to be documented and publicized to a greater extent than often has been the case.

FSD can help Identify Appropriate Development Strategies

On-farm systems studies were carried out in village areas south of Bangalore, India to determine the appropriateness of the current livestock policy.

Farmers without irrigated crop land are the poorest in this area and at greatest risk from the effects of climatic and biological variations.

These farmers depend on small livestock herds for cash income and on subsistence cropping for home consumption. When their small-scale livestock production fails, farmers are forced to seek outside employment for cash, and thus cropping activities suffer.

This farming system study identified two improvements that would contribute to better livestock productivity: control of epizootics and a 'preventative veterinary medicine service' focused on improved fertility management among small herd-owners.

Previously, veterinary service policy had been to target larger herds where the immediate payoff had appeared to be much greater. Such a livestock policy change should lead to greater stability in the production of animal products for local markets; less reliance on seasonal employment and, therefore, greater stability of the household labour force; and more stable subsistence cropping activities.

Inter-household Relations and Trade-offs: An Intragenerational Issue

In Botswana, farmers have a tradition of sharing draught power, In this semi-arid environment, field operations such as ploughing and row planting depend on access to draught in a timely fashion (i.e., days on which there is adequate soil moisture for ploughing/planting). If a technology system is introduced that requires more draught, rather than less, it may mean that the household controlling the draught may not be able to routinely allow other households to also use their animals under a sharing arrangement. Thus, helping one household may harm other households in the community.

Intra-household Decision Making and Trade-offs: Another Intra-generational Issue

An FSD programme in Amphoe Pharao, Thailand discovered a positive and unexpected benefit from interaction at the intra-household level. The introduction of improved and more commercially oriented poultry operations was targeted towards women members of households, because they traditionally manage and derive income from poultry.

However, improved technological interventions not only required new inputs by men but were discovered to provide profitable use of under-employed male labour in many households.

This example demonstrates decision making with interactions between members of the household that are desirable for all: a win-win situation. Often, however, decisions for change favour one segment of the household, while adversely impacting on another.

Some Key Concepts Important in the Farming Systems Approach to Development

Definition of a Farming System

To design appropriate or relevant ways of helping farmers, it is essential to understand the conditions under which farmers are operating. They have in fact, very complicated farming systems. The factors that have an influence on what the farming system will be. The operator of the farming system is the farmer or the farming family. To farmers, the way in which they earn their living and the economic, social, and cultural well-being of their households are linked closely and cannot be separated.

The members of the farming household have three basic types of inputs: land, labour, and capital. Management involves allocating these to three different activities or processes, that is, crops, livestock, and off-farm enterprises. In making decisions on how to allocate their inputs in producing one or more products, farmers have to make some difficult decisions.

These decisions will involve using their knowledge to come as close as possible to fulfilling the goal(s) for which they are striving. These goal(s) may vary from farmer to farmer (e.g., maximizing their income, producing enough food to feed the family, etc.). Many farmers are likely to want to maximize their incomes, once they have made sure they are producing enough food to feed their family and have met other societal obligations. The resulting combination of products (i.e., crops, livestock, and off-farm enterprises) they are producing with their inputs results from the farming system they have adopted. Nevertheless, the extent to which the farming system fulfils the goal(s) they have chosen will depend on the managerial skills of the farming family and its ability to make good decisions in the allocation of inputs in a very uncertain production environment.

However, some parts of the environment that influence what the farming system will be are outside the control of the individual farming family thus causing uncertainty as far as the farmer is concerned. The 'total' environment in which farming households operate consists of two parts: the technical (i.e., natural or physical) element and the human element.

- The technical element determines the types of, and physical potential of; livestock and crop enterprises. For example, in some areas, the low level of rainfall allows sorghum, but not maize, to be grown on rainfed land. The technical element includes physical and biological factors. These often are modified to some extent through technology developments — for example, increasing water availability through irrigation Improving soil quality by adding fertilizer, breeding for yield stability during drought, etc.
- The farming system that actually evolves, however, is only part of what is potentially possible. The human element is important in determining what the actual farming system will be. The human element consist of two types of factors: exogenous and endogenous.

Exogenous factors (i.e., the social environment) are largely out of the control of the individual farming family. These factors will influence what the farming family can do and can be divided into three broad groups:

- Community structures, norms, and beliefs.
- External institutions, which include extension, credit, and input distribution systems on the input side and markets on the output side.
- Other influences, such as population density, location, and infrastructure.

Endogenous factors, on the other hand, are those the individual farming household controls to some degree. These include the types of inputs mentioned earlier, that is, land, Iabour, and capita]. It is important to recognize that these resources and managerial ability vary among households and regions. The resources vary on the basis of quantity and quality, both of which influence the performance and the potential of the system. In addition, these inputs or resources may or may not be owned by the household. Access to one or more of these resources may be on another basis of use (e.g., borrowing draught animals), which may limit or restrict the ease or intensity of use and thus, in turn, affect the goals and performance of the farm family.

Nevertheless, it is the fanning family that decides on the farming system that will emerge. However, this system will be influenced and sometimes constrained by the technical element and exogenous factors. The farming system is obviously complex, and the results can vary greatly because of differences in the 'total, environment. These facts help explain why some technology thought to be relevant often has not been adopted, or when it has, why the degree of adoption has varied widely. Not considering the human element in agricultural research has contributed to many socalled 'improved' technologies being irrelevant.

The Limited-resource Household: Status, Goals, and Appropriate Technology

Typical limited resource farm households in low income countries generally are characterized as follows:

"Farm households, with access to their means of livelihood in land, utilizing mainly labour in farm production, always located in a larger economic system, but fundamentally characterized by partial engagement in markets which tend to function with a high degree of imperfection."

Important points to note about this definition are that:

- Such households often are subordinate to some other external forces. That is, they do not have complete control over their own destiny and are in a process of transition. This is because of not being integrated fully in the market economy.
- They have access to land to pursue their livelihood. They cultivate the land largely with family labour, in conjunction with only small amounts of capital.
- The input or factor markets (i.e., land, labour, and capital) work poorly. Concerning land, non-market rights of access or non-price forms of tenure are more likely to operate than a freehold market. The capital markets usually are developed poorly and variable production inputs are often not available. As a result:

  "Credit and interest rates may be tied to other factor prices like land and labour within a dependent economic relationship. Thus factor markets may be locked together contractually rather than being independent."

Also, market information may be highly imperfect (i.e., not available or readily accessible). Thus, it is not surprising that sharing and reciprocity often exist between households. In these non-market transactions, exchange of unlike goods and services take place, which are not valued by market prices. Sometimes, such relationships exist between households that are related to each other.

- Another reason why such households are integrated only partially into the market economy is because they consume a proportion of the product they produce. This enables them to have some ability to survive independently of the larger system, which may be important in explaining their economic behaviour.

  A number of microeconomic theories have been developed in an effort to explain the economic behaviour of farm households. These are based on various assumptions about household goals and the characteristics of the markets within which households make their decisions. Ellis [1988] notes that these theories are not mutually exclusive, sharing as they do certain key assumptions such as:

- The household is a single decision-making unit for economic analysis purposes. As a result, it maximizes a single utility (satisfaction) function that represents the joint welfare of its members.
- Profit maximization and utility maximization coincide where income is the only variable in the utility function. Profit maximization is always one of the components of utility maximization" when all the input and output markets are fully formed and competitive.

Ellis [1988] suggests that the theories predict different results because of different assumptions about the working of the factor and product markets, rather than because of differences in assumptions about household goal(s). In fact, an important characteristic distinguishing many of the theories results from different assumptions about labour markets and the allocation of household labour time, In support of these assertions, Ellis [1988] then discusses a number of these theories of economic behaviour. He calls these profit maximization, risk aversion, drudgery aversion, farm household theories, and share tenancy. In doing so, he recognizes the impossibility of separating short-run household decisions from the wider social relations of production. Consequently, there is likely to be a certain universality in the aspirations of limited-resource farming households. The most important elements are income, effort avoidance, and risk avoidance. In economic terms, this means such households try to increase their utility or satisfaction, which increases with income but decreases with greater effort or higher levels of risk. This can be restated as maximizing income for any given level of effort or risk. Alternatively, it can mean reducing risk, or effort, for a given level of income.

Attempts to maximize utility or satisfaction take place within a set of constraints. As implied above, differences in the constraints, rather than the aspirations, lead to the most important differences in farming systems. As discussed earlier, indicates some of the determinants of the farming system. It is easy to visualize constraints relating to the technical element causing differences in the type and productivity of farming systems, However, also important in the differentiation process are the factors relating to the human element. The quantity, quality, and ratios (i.e., particularly between land and labour) of the endogenous factors (i.e., factors of production — land, labour, and capital) play an important role in further differentiating farming systems. The extent to which they can be modified will depend on the degree of integration into the factor markets (i.e.? input markets). Also the products resulting will be influenced to some extent by community norms and beliefs and the degree of integration into the product market. The above discussion has important implications for what is appropriate technological change. For example, limited resource households often are likely to be very little above survival level and, hence, will place a very high premium on improved stability of production.

The levels, qualities, and relative proportions of the factors of production are important in indicating the most appropriate route for improving the productivity and profit of the farming system — which is maximizing the return to the most limiting factor. For example, in areas of high land/labour ratios (i.e., low population densities), labour saving (e.g., mechanical, extensive livestock production) types of technology/systems are likely to be most appropriate. In contrast, in areas of low land/resident ratios (i.e., high population density) land saving technologies/systems (e.g., biological and some chemical, intensive livestock production) will likely be more appropriate. However, increasing the return to the most limiting factor can have an indirect positive or negative impact on the use and productivity of other inputs. Thus, in view of the imperfect operation of the factor markets, care has to be taken in evaluating what would be the most relevant technologies.

For example, in a semi-arid area, initial superficial examination may indicate that because of the high population densities, strategies designed to increase the productivity of land would be most important. However, given the seasonal nature of agriculture in semi-arid areas, there will be periods of intense labour activity during the year. These will create labour bottleneck periods, followed by periods of underemployment. Thus, land-augmenting strategies using labour at critical labour bottleneck periods — when the opportunity cost of labour is high (i.e., there are good alternative uses for labour that can yield an equally high return) — could be unattractive to the household. This is likely if adoption of such strategies involves the use of more labour at such labour bottleneck periods. It will also be likely if the productivity per unit of labour applied at that time is higher in alternative uses.

The results indicated the superiority of the improved maize technology over both improved or indigenous practices for sorghum and cotton in terms of both return per unit of land and labour. Note that for both the sorghum and cotton improved technologies, the return per unit area was higher than that from employing indigenous practices; but that the return per unit of labour used during the June/ July period was the same or lower. Unlike the sorghum and cotton improved technologies, the improved maize technology was, from a profit viewpoint, suitable for farmers whether they were faced with a labour or land limitation.

Not surprisingly, since the 1970s, improved maize technologies have been adopted widely in Northern Nigeria, in a mini-Green Revolution. Although, initially the prime motivation for growing maize was as a cash crop, it eventually became a more important dietary item as farming households made trade-off decisions between growing the high yielding maize rather than the low yielding sorghum.

Equitability Considerations-Intra-and Inter-Generational

Intra-Household or Gender Issues

Many development projects have assumed that the male head-of-household provides the majority of agricultural labour and is the sole decision maker concerning agricultural activities. This approach fails to recognize the important contribution to women and children to agricultural production in many parts of the world.

Although data are very unreliable, the indications are that female family labour ranges from 2 to over 50 percent of farm labour. Women's participation in production agriculture may range from zero in some Moslem countries to the situation in Africa south of the Sahara, which is a region of 'female farming systems' where women may manage and work more than one-third of the farms. Despite the great contribution made by women, development projects have been slow to recognize this contribution. As Jacobson points out:

> *"The reason is brutally simple: women perform the lion's share of work in subsistence economies, toiling longer hours and contributing more to family income than men do. Yet in a world where economic value is computed in monetary terms alone, women's work is not counted as economically productive when no money changes hands."*

She goes on to suggest that this gender bias is compounded in many societies by discrimination based on class, cast, or race. This situation can lead to complications because of differentiation in the types of labour available from household sources, Microeconomic theories that make simplistic assumptions about farming households add to the problem, For example, they assume a single decision-making unit and a single utility function representing the joint welfare of family members. However, increasingly this has been recognized as a problem, because intra-household relationships can have a major impact on economic behaviour, Some important points to note about intrahousehold relationships are as follows:

- Such relationships are of particular importance in situations where there is a high degree of congruence between the production and consumption units. They are also important where there is some imperfection in terms of integration with the factor and product markets.
- The economic situation of individuals within households is influenced by the social norms or behaviour of the society within which they reside, Therefore, specific economic roles are socially assigned to men, women, and children, Also, division of property, labour and income within households is socially, and not biologically, determined. As a result, males and females within households sometimes may have different income streams, with different sources and different destinations.
- Thus, it is wrong to assume that male and female labour can substitute for each other across the range of household activities. As a result, rigidities in time allocation may reduce the ability of households to respond to market signals. Therefore? such division of labour can impose constraints on seasonal labour inputs and ability to respond to price changes. As a result, it can have a major impact on the composition and level of farm output.
- Because of the above, increases in household income are unlikely to benefit all members equally. This often is exacerbated by the social subordination of women in many societies. As pointed out above, agricultural policies and projects often have ignored the role of women in agricultural production — not only as operators but also as decision makers. As a result, it is not surprising that technological innovations have tended to disadvantage women relative to men.

Therefore, an understanding of operational constraints facing limited resource households, and the relationships that exist between members of the households and between households is critically important in determining what are likely to be appropriate criteria for evaluating potential technologies. In connection with this, it is also important to understand the farming household members' goals and incentives, farming and non-farming activities, and resources available.

Inter-Generational or Sustainability Issues

As was pointed out previously, incorporating equity issues relating to the intergenerational impact of development is one of the challenges facing FSD.

There is a great deal of concern about the impact of an ever increasing world population on the world's natural resources. This concern was evident at the Earth Summit held in June 1993 in Rio de Janeiro. Part of this concern was related to the loss of nearly 500 billion tons of topsoil through erosion in the 20 years between the previous meeting in Stockholm and the Rio de Janeiro meeting. According to Brown [1993], this and other environmental factors have contributed to a decrease in the rate with which food output is growing — an alarming situation given the increasing population to be fed. Most farmers, and hence development programs, are concerned with local short-run production problems rather than the broad picture presented at the Earth Summit. This its logical, given the farmer's need to survive at the present, but it does not excuse the FSD team from considering the long-term, intergenerational and sustainability, impacts of their programmes, Although not the only factor to consider in the inter-generational and sustainability area, a major concern is soil conservation," Norman and Douglas [1994] identify a number of concerns when addressing soil conservation issues.

- Good land husbandry requires a mix of improved farming practices integrated with farm policies and programmes. Thus, it is essential to understand the current farming system in order to design effective conservation measures. This implies that a systems perspective be used.
- The goal should be to prevent rather than cure soil erosion. This means that the causes of soil erosion must be understood and treated, rather than concentrating on treating the symptoms.
- To be most effective, conservation programs should be planned with the full knowledge and participation of the farmers. FSD team members must be able to communicate effectively with

farmers to identify their ideas, perceptions, and priorities, so these can be taken into account in developing strategies for improvement.

- If farmers are going to implement changes, they must be convinced that they can obtain short-term benefits from change (i.e., have an incentive to alter present practices). Thus, it is important to try and identify strategies that will have short-term benefits as well as provide long-run conservation improvements. When dual-purpose strategies cannot be designed, the long-term conservation activities should be linked with measures that provide short-term production/welfare benefits.
- For any conservation activities that require community action (e.g., soil conservation on a watershed), it is important that the communities have responsibility and control and also reap the benefits of such action, in an equitable manner.
- Long term programmes are more effective than short-term programmes. Thus, it follows that the change will take place over a long period, necessitating adjustments and iterations in programmes.

Balancing Individual Farm and Community Objectives

Rapid changes are occurring in cropping system recommendations and practices in the Andes of Bolivia. Some of these changes are in conflict with objectives and beliefs held by the community. Individual villagers in this region occupy field sites for two or three years within communal lands and then shift to new sites. Individuals make their own decisions about cropping practices, and the community collectively decides when shifts are in order. The fallow period typically lasts from 3-15 years and provides communal grazing, rejuvenates soil fertility, aids with pest control in crop production, and provides fuel for the household. Furthermore, the long-term fallow system, aynuga, contributes symbolically to the well-being of the community and its sense of stewardship.

Adoption of 'improved, cropping practices by individual farmers on their fields could change the way in which they wish to occupy space in the long-term fallow system. Solutions to this restraint will require balancing objectives of individual farmers and community. These considerations are in addition to technical considerations in soil conservation and imply that the FSD team must possess skills other than just technical skills. This is implied in the example given in Box 4.2. Although this discussion has centred on soil conservation, the approach and factors to consider can he applied equally well to other programme areas where there are inter-generational or sustainability considerations-which will include most programme areas.

Recommendation Domains

A recommendation is information that farmers can use to improve the productivity of their resources (i.e., farming system). A good or successful recommendation, therefore, would consist of the practices that farmers would follow, given their current resources. If they had all the information actually available to the researchers. Obviously, the development of recommendations for farmers should be as efficient as possible.

Because it is impossible to make a separate recommendation for each farmer practical compromises have to be made. As indicated earlier, this is done by stratifying the farmers into groups as homogeneous as possible. Farmers in these groups should have similar circumstances, resources, problems, and solutions to those problems. In this situation, therefore, it is usually assumed that the same recommendation will be suitable for all the farmers in the group. Such a group of farmers is a particular recommendation domain. Recommendation domains may be defined by agro-ecological (technical) and/or by socio-economic (human) circumstances. The definition of the recommendation domain depends on the particular recommendation.

Therefore, by dividing farmers into homogeneous groups or recommendation domains, one is recognizing:

- The heterogeneity that exists in the farming population.
- The likelihood that farmers, within a group, will have similar problems.
- The increased probability that farmers within a group, will be interested in the same potential solution to the identified problem(s).

Thus, the purpose of the groups is to highlight similarities within the recommendation domains and the differences between and among the recommendation domains. Similarities and differences are important only with respect to the objectives: increasing agricultural productivity by establishing experiments and generating recommendations. Because farm families differ from one another, it is important to distinguish the characteristics on which they vary, how those variables relate to the problems they have, and the usefulness of solutions to resolving those problems.

In the descriptive/diagnostic exercise, as will be seen later, one of the purposes is to tentatively define the potential recommendation domains, although these may need to be adjusted in the light of later experience.

An interesting short-cut that does not require a very precise definition of recommendation domains is the use of research-oriented farmer groups.

These are discussed in a later section, Briefly, this involves offering to the farmers a number of technologies, from which they select those they are interested in testing. The characterization of the group selecting the option provides a quick indication of the possible recommendation domain for the proposed technology and reduces the chance of possible errors in the definition of the recommendation domains.

Participatory Research

During recent years, the farming systems approach has shifted dramatically from an emphasis on farmer based but researcher managed studies towards methods that stress farmer participation, or say, in the research process. The question asked is 'who is best positioned to lead an inquiry into a farming system'?'. Farmer-based, researcher-managed studies are necessary when:

> *"Farmers may not have adequate technical explanations of their problems nor know the range of opportunities for improving their conditions, but learning more about farmers helps the researchers produce better technologies and extension workers promote FSR results more effectively."*

However, participatory techniques have proven very useful in allowing farmers to express themselves on a wide range of issues. In the extreme, the purposes of a participatory orientation could be simply to organize a voice for farmer concerns and to impact on political decisions. Another focused objective might be to establish nearly autonomous farmer groups that can be self-motivated and self-directed in problem-solving investigations, But, in practice, most participatory activities involve some balance, or a partnership, between researcher management and farmer say in the research dialogue.

The justification for shifting towards participatory techniques lies in the need to give farmers or groups of farmers the freedom to conceptualize and express an accurate model of how their farming system works, Participants can synthesize and exploit a comprehensive farm model in terms of systems structures, processes, and farm objectives certainly more efficiently, and often more effectively, than can FSD teams who study the system from the outside.

When embarking on greater farmer participation, FSD can choose to invite farmers to manage or comanage all or only some stages of the research effort, These stages might include: descriptions of the system, diagnoses of constraints, designs of new options, pre-screening or other assessments of options, selections of options for actual testing, reporting assessments of options to FSD, and even taking charge of disseminating useful information to other farmers.

The role of FSD team members in participatory research usually is not reduced. In tact, it may require a greater awareness of issues (i.e., farmers sensitivities, social customs, and technical understanding) than in more researcher managed work. In participatory research, FSD members should search out ways to:

- Facilitate the articulation of the participants' positions— often consensus of a group.
- Present hypotheses for discussion or testing.
- Link discussions with information from secondary sources that is not being taken into consideration by the participants.
- Communicate findings to outside agents.
- Link findings to the larger agricultural research and development context.
- Bring outside decision makers (e.g., research, extension, development, and planning agencies) directly into the participatory process.

The correct balance between participatory methods and farmer-based, researcher-managed methods must be determined by the FSD team based on what is appropriate for each activity. The choice should be for the method or combination of methods that will give the greatest problem-solving efficiency. The choices, especially when a large degree of uncertainty exists about the issues and about the types of responses that can be anticipated, might be arrived at through an iterative process in which FSD teams routinely re-evaluate the techniques in their work plan.

Farmer participation is discussed or implicit under a number of the specific methodology sections of this manual.

Obviously, a strong participatory role occurs in informal survey work and more particularly for PRA in Section 8.4. A range of participation in trial management is discussed with emphasis on how this participation corresponds to the objectives of the trial. A participatory mode is greatest in FMFI trials. Farmer groups, listed under trial management methodology, are actually useful participatory forums to address essentially all stages of the research effort.

Bibliography

Arthur W. Gilbert, Mortier F. Barrus and Daniel Dean: *Growing and Breeding of Potatoes*, Asiatic Pub, Delhi, 2006.

Ausubel, F.M. : *Current Protocols in Molecular Biology,* New York: John Wiley and Sons, 1989.

Banki, L.: *Bioassay of Pesticides in the Laboratory,* Akademiai Kiado, Budapest, 1978.

Barbeau G.: *Tropical Fruits in Nicaragua*, Managua, Nicaragua Ministerio de Desarrollo Agropecuario, Agraria, 1990.

Chadha K. L. and Pareek O. P.: *Advances in Horticulture: Fruit Crops,* New Delhi, Malhotra Publishing House, 1993.

Chaturvedi, O.P. : *Agroforestry : Natural Resource Sustainability, Livelihood and Climate Moderation*, Satish Serial, Delhi, 2009.

Divya, M P : *Social Forestry and Agroforestry*, Satish Serial Publishing House, Delhi, 2008.

Ferentinos L.: *Proceeding of the Sustainable Taro Culture for the Pacific Conference*, Honolulu, HITAHR, 1993.

Gupta, J.P. and B.M. Sharma: *Agroforestry for Sustained Productivity in Arid Regions*, Scientific, Delhi, 1997.

Herbert Meschkowski: *Ways of Thought of Great Mathematicians*, Holden-Day Inc., San Francisco, 1964.

J S Amarnath and A P V Samvel: *Agri-Business Management*, Satish Serial Pub, Delhi, 2008.

Jha Timir Baran and Ghosh Biswajit : *Plant Tissue Culture : Basic and Applied*, Universities Press, Delhi, 2005.

Kapoor, R.L. and M.L. Saini: *Plant Breeding and Crop Improvement*, CBS, Delhi, 1997.

Kass, Leon R. : *Life, Liberty, and the Defense of Dignity: The Challenge for Bioethics*, San Francisco: Encounter, 2002.

Kunelius T.: *Annual Ryegrasses in Atlantic Canada,* Ottawa, Agriculture Canada, 1991.

Lindee, M.S. : *The DNA Mystique: The Gene as a Cultural Icon*. New York: W.H. Freeman, 1995.

Mal, Pratap : *Infrastructural Development for Agriculture and Rural Development*, Mohit, Delhi, 2001.

Montagnini, Florencia : *Environmental Services of Agroforestry Systems*, International Book Distributing Co, Delhi, 2007.

Nair, P K Ramachandran : *An Introduction to Agroforestry*, Springer, Delhi, 2008.

Nobel, P. S.: *Physicochemical and Environmental Plant Physiology*, Academic Press, San Diego, 1999.

Oldham P.: *Cost of Production of Major Tree Crops in Dominica*, Roseau, Ministry of Agriculture, 1991.

Pathak, P.S. and Ram Newaj: *Agroforestry : Potentials and Opportunities*, Agrobios, Delhi, 2003.

Politycka, B and C L Goswami: *Plant Physiology : Research Methods*, Scientific, Delhi, 2007.

Puri, Sunil and Pankaj Panwar: *Agroforestry : Systems and Practices*, New India Pub, Delhi, 2007.

Qystein V. Sjaastad: *Physiology of Domestic Animals*, International Book Distributing Co., Delhi, 2005.

Ragone D.: *Breadfruit: Artocarpus Altilis (Parkinson) Fosberg*, Rome, International Plant Genetic Resources Institute, 1997.

Rao, S. Balaachandra: *Indian Mathematics and Astronomy*, Bangalore, Jnana Deep Publications, 1994.

Savindra Singh, H.S. Sharma and Sunil Kumar De: *Geomorphology and Environment*, ACB Pub, Delhi, 2004.

Sen, N.L. : *Manual of Agroforestry and Social Forestry*, Agrotech, Delhi, 2004.

Tyagi, I.D. : *Plant Breeding and Genetics at a Glance*, South Asian, Delhi, 2005.

Umrani, Ramesh and C.K. Jain: *Agroforestry : Systems and Practices*, Oxford Book Company, Delhi, 2010.

Vanderplank, J.E.: *Plant Diseases: Epidemics and Control,* New York, NY, USA, Academic Press, 1963.

White, G.F.: *Natural Hazards: Local, National, Global*, Oxford University Press, New York, 1974.

William H. Elliott and Daphne C. : *Biochemistry and Molecular Biology*, Oxford University Press, Delhi, 2005.

Yadav, M : *Nutritional Biochemistry and Metabolism,* Arise Pub, Delhi, 2008.

Index

I

L

M

P

R

S

T

V

❑❑❑